Integrated Urban Water Arid and Semi-Arid Reg

Urban Water Series – UNESCO-IHP

ISSN 1749-0790

Series Editors:

Čedo Maksimović
Department of Civil and Environmental Engineering
Imperial College
London, United Kingdom

Alberto Tejada-Guibert
International Hydrological Programme (IHP)
United Nations Educational, Scientific and Cultural Organization (UNESCO)
Paris, France

Integrated Urban Water Management: Arid and Semi-Arid Regions

Edited by

Larry W. Mays

CRC Press is an imprint of the
Taylor & Francis Group, an **informa** business
A BALKEMA BOOK

Cover illustration

Central Arizona Project aqueduct through residential area in Scottsdale, Arizona – US Bureau of Reclamation, with kind permission from US Bureau of Reclamation.

Published jointly by

The United Nations Educational, Scientific and Cultural Organization (UNESCO)
7, place de Fontenoy
75007 Paris, France
www.unesco.org/publishing

and

Taylor & Francis The Netherlands
P.O. Box 447
2300 AK Leiden, The Netherlands
www.taylorandfrancis.com – www.balkema.nl – www.crcpress.com
Taylor & Francis is an imprint of the Taylor & Francis Group, an informa business, London, United Kingdom.

Typeset by Macmillan Publishing Solutions, Chennai, India
Printed and bound in Hungary by Uniprint International (a member of the Giethoorn Media-group), Székesfehévár.

ISBN UNESCO, paperback: 978-92-3-104061-0
ISBN Taylor & Francis, hardback: 978-0-415-45348-6
ISBN Taylor & Francis, paperback: 978-0-415-45349-3
ISBN Taylor & Francis e-book: 978-0-203-89544-3

Urban Water Series: ISSN 1749-0790

Volume 3

British Library Cataloguing in Publication Data
A catalogue record for this book is available from the British Library

Library of Congress Cataloging-in-Publication Data

Integrated urban water management : arid and semi-arid (asa) regions / edited by Larry W. Mays.
p. cm. – (Urban water series ; v.3)
Includes bibliographical references and index.
ISBN 978-0-415-45348-6 (hardcover : alk. paper) – ISBN 978-0-415-45349-3 (pbk. : alk. paper) – ISBN 978-0-203-89544-3 (e-book) 1. Municipal water supply–Arid regions–Management–Case studies. 2. Water quality management–Arid regions–Case studies. 3. Water-supply–Arid regions–Management–Case studies. 4. Groundwater–Arid regions–Management–Case studies. 5. Arid regions–Environmental conditions–Case studies. I. Mays, Larry W. II. Title.
TD220.2.I58 2009
363.6'1091732–dc22 2009006423

Foreword

Water scarcity has become a major global concern, with one-third of the world facing water shortages due to rapid population growth and deteriorating water quality. In particular, many cities around the world have been experiencing water shortages and scarcity, which are expected to increase further due to greater water demand caused by population growth, rapid urbanization and economic growth. Effects of climate change and variability, and poor management of water resources often exacerbate the problem. In arid and semi-arid regions throughout the world, this can be accentuated by prolonged droughts reflecting climate change and the process of desertification. Because of the highly specific characteristics of their climate, cities in these regions will probably face major water challenges in the near future.

With the aim to contribute to addressing these challenges, this book examines the integrated management of water resources in urban environments in arid and semi-arid regions around the world, focusing on its specificities. It addresses a broad range of issues, such as how growing water demand could be met given the scarce water resources of the region and how urbanization impacts water resources management and the urban water cycle in arid and semi-arid regions.

This book is one of the main outputs of the project on "Integrated Urban Water Management in Specific Climates", implemented during the Sixth Phase of UNESCO's International Hydrological Programme (IHP, 2002–2007), and represents the result of efforts and deliberations of experts whose contributions also enriched the publication with case studies from across the world emphasizing specific needs and challenges of urban water management in arid and semi-arid regions worldwide. The contribution of Mr Larry W. Mays as the lead editor of the book was significant and gratefully acknowledged. This publication, which is part of the UNESCO-IHP Urban Water Series, was prepared under the responsibility and coordination of Mr J. Alberto Tejada-Guibert, Deputy-Secretary of IHP and Responsible Officer for the Urban Water Management Programme of IHP, and Ms Sarantuyaa Zandaryaa, Programme Specialist in urban water management and water quality at IHP. We extend our thanks to all the contributors for their remarkable effort, and we are confident that the conclusions, recommendations and case studies presented in this volume will prove to be of value to urban water management practitioners, policy- and decision-makers and educators alike throughout the world.

András Szöllösi-Nagy
Secretary of UNESCO's International Hydrological Programme (IHP)
Director of UNESCO's Division of Water Sciences
Deputy Assistant Director-General for the Natural Sciences Sector of UNESCO

Preface

This book examines the integrated management of water resources in urban environments in arid and semi-arid regions around the world. The UNESCO-IHP has grouped water-related environmental problems faced by cities into the following broad categories:

- access to water and sanitation infrastructure and services
- urban wastewater pollution
- resource degradation
- water-related hazards.

Cities in developing countries typically face all four problems. In many cases, all four problems occur simultaneously with high intensity over long periods. As urban populations continue to grow rapidly around the world with the addition of many new megacities, this urbanization process creates many challenges for the development and management of water supply and water excess management systems. Water scarcity is certainly one of the many challenges in urban areas in arid and semi-arid regions of the world, particularly with the rising level of urbanization.

The first six chapters give an overview of the various aspects of integrated urban water management in arid and semi-arid regions. The urban water system is considered as a single integrated whole. Urban water management is considered herein as consisting of two major entities, urban water supply and urban water excess management systems. Urban water supply systems include all the system components to provide drinking water and distribute it to users in addition to all the system components needed to collect and treat the wastewaters. Special emphasis is given to technologies, such as artificial recharge, water transfers, desalination, and harvesting of rainfall, which are typically not part of the conventional urban water supply systems, but are used in arid and semi-arid regions as viable sources of water supply. Water excess management systems include both the stormwater management system and the floodplain management system. In the big picture, integrated water management must include:

- The systematic consideration of the various dimensions of water: surface and groundwater, quality and quantity.
- The implication that while water is a system it is also a component which interacts with other systems.
- The interrelationships between water and social and economic development.

Water resources sustainability must be a major overall goal of water management. Herein water resources sustainability is defined as the ability to use water in sufficient quantities and quality, from the local to the global scale, to meet the needs of humans and ecosystems for the present and the future to sustain life, and to protect humans from the damages brought about by natural and human-caused disasters that affect the sustaining of life. Unfortunately, many urban areas of the world, particularly in developing parts of the world, are unsustainable from the viewpoint of water.

Case studies for both developed and developing regions of the world are presented in order to emphasize the various needs and challenges of urban water management in arid and semi-arid locations around the world. These case studies include: Mexico City, Mexico; Tucson, Arizona; Awash River Basin, Ethiopia; China; and Cairo, Egypt.

With all the uncertainties that exist, there is one certainty: poor water management hurts the poor most.

Contents

List of Figures

Case Studies

List of Tables

Acronyms

ABWRAA	Awash Basin Water Resources Administration Agency
ADWR	Arizona Department of Water Resources
AFRP	Agua Fria Recharge Project
AMA	Active Management Area
AOC	Assimilable Organic Carbon
APT	Advanced Primary Treatment
ASA	Arid and Semi-Arid
AWBA	Arizona Water Banking Authority
AWS	Assured Water Supply
BAC	Biological Activated Carbon
CAP	Central Arizona Project
CAWCD	Central Arizona Water Conservation District
CAVSARP	Central Avra Valley Storage and Recovery Project
COD	Chemical Oxygen Demand
CPWC	Cooperative Programme on Water and Climate
DBP	Disinfection By-Product
DSS	Decision Support System
DWC	Dubai World Central
EEPCO	Ethiopian Electric Power Corporation
ENSO	El Nino Southern Oscillation
EU	European Union
FEMA	Federal Emergency Management Agency
FCDMC	Flood Control District of Maricopa County
FLOW	Friends of Lower Olentangy Watershed
FWID	Flowing Wells Irrigation District
GCC	Gulf Cooperation Council
GCM	Global Circulation Model
GDP	Gross Domestic Product
GIS	Geographical Information Systems
GMA	Groundwater Management Act
GMC	Groundwater Management Code
GMMRP	Great Man-Made River Project
GRUSP	Granite Reef Underground Storage Project
ICM	Integrated Catchment Management

IGA	Inter-Government Agreements
INA	Irrigation Nonexpansion Area
IPCC	Intergovernmental Panel on Climate Change
IWA	International Water Association
IUWM	Integrated Urban Water Management
IWRM	Integrated Water Resources Management
LNSO	La Nina Southern Oscillation
MAR	Managed Aquifer Recharge
MDG	Millennium Development Goals
MENA	Middle East and North Africa
MDWID	Metropolitan Domestic Water Improvement District
MIS	Management Information Systems
NFIP	National Flood Insurance Program
NOM	Natural Organic Matter
PAC	Powdered Activated Carbon
PPC	Permanganate Composite Chemical
RCUWM	Regional Centre on Urban Water Management
RS	Remote Sensing
RTU	Remote Terminal Unit
SAT	Soil-Aquifer Treatment
SAVSARP	Southern Avra Valley Storage and Recovery Project
SAWRSA	Southern Arizona Water Rights Settlement Act of 1982
SCADA	Supervisory Control Automated Data Acquisition
SRP	Salt River Project
SST	Sea Surface Temperature
SS	Suspended Solids
TAMA	Tucson Active Management Area
TDS	Total Dissolved Solids
TOC	Total Organic Carbon
TSS	Total Suspended Solids
UNDP	United Nations Development Program
USAID	United States Agency for International Development
US EPA	United States Environmental Protection Agency
USBR	United States Bureau of Reclamation
US$	US dollars
WHO	World Health Organization
WSSD	World Summit on Sustainable Development
WSM DSS	Water Strategy Man decision support system
WT	Water Treatment
WWT	Wastewater Treatment

Glossary

Adiabatic cooling A natural atmospheric process whereby an air mass cools due to lower pressures as it rises, while maintaining the same volume. This effect can cause water vapor to condense and form rain or snow in the presence of condensation nuclei.

Aeration Addition of air to water resulting in a rise of its dissolved oxygen level.

Aggradation Process of raising a land surface by the deposition of sediment.

Alfisols Mineral soils that have umbric or ochric epipedons, argillic horizons, and that hold water at less than 1.5 MPa tension during at least 90 days when the soil is warm enough for plants to grow outdoors.

Alluvial channel Channel with a movable bed in loose sedimentary materials.

Anaerobic condition Condition of water in which the dissolved oxygen is too low to support aerobic bacteria.

Anticyclone The rotatory outward flow of air from an atmospheric area of high pressure; also, the whole system of high pressure and outward flow.

Aquifer Permeable water-bearing formation capable of yielding exploitable quantities of water.

Aquitard; *syn.* semi-confining bed; Geological formation of a rather impervious and semi-confining nature which transmits water at a very slow rate compared with an aquifer.

Aridisols Mineral soils that have an aridic moisture regime, an ochric epipedon, and other pedogenic horizons but no oxic horizon.

Artesian well Well tapping a confined or artesian aquifer in which the static water level stands above the surface of the ground.

Basin Drainage area of a stream, river or lake.

Biochemical Oxygen Demand (BOD) The amount of dissolved oxygen consumed by micro-organisms as they decompose organic material in polluted water; a water quality indicator; BOD5 is the biochemical oxygen demand over a five-day period, that is, the amount of oxygen consumed by micro-organisms over a five-day period as they decompose organic matter in polluted water.

Biofiltration Combined physical and chemical processes of filtration and adsorption with the uptake and processing of nutrients by attached micro-organisms.

Biorention area A vegetated surface depression designed to collect, store and infiltrate runoff; where needed, the underlying soil layer is replaced with bioengineered soil.

Brine Very concentrated salt solution (conventionally above 100 000 mg/l) often produced by evaporation or freezing of sea water.

Coagulation process Process of adding a chemical (the coagulant) which causes the destabilization and aggregation of dispersed colloidal material into flocs.

Combined Sewer Overflow (CSO) Discharge of a mixture of stormwater and domestic waste when the flow capacity of a sewer system is exceeded during rainstorms.

Convective rainfall Precipitation caused by convective motion in the atmosphere.

Drainage basin; syn. watershed, catchment; The land area drained by a river or a body of water.

Drywell A bored, drilled, or driven shaft or a dug hole or subsurface fluid distribution system, whose depth is greater than its largest surface dimension, which is completed above the water table so that its bottom and sides are typically dry except when receiving fluids.

El Niño Southern Oscillation (ENSO) The anomalous appearance, every few years, of unusually warm ocean conditions along the tropical west coast of South America. This event is associated with adverse effects on fishing, agriculture, and local weather from Ecuador to Chile and with far-field climatic anomalies in the equatorial Pacific and occasionally in Asia and North America as well. (EB) El Niño is opposite to La Niña.

Electrodialysis A process that uses electrical current applied to permeable membranes to remove minerals from water. It is often used to desalinize salty or brackish water.

Endocrine Disrupter Compounds (EDCs) Chemicals that interfere with the normal function of hormones and the way hormones control growth, metabolism and body functions.

Endorreic Draining into interior basins.

Enteric diseases Diseases of or relating to the small intestine.

Entisols Mineral soils that have no distinct subsurface diagnostic horizons within 1 m of the soil surface.

Erratic pressure Pressure liable to sudden unpredictable change.

***Escherichia coli* (E. coli)** Bacteria present in the intestine and feces of warm-blooded animals. E. coli are a member species of the fecal coliform group of indicator bacteria. Their concentrations are expressed as number of colonies per 100 ml of sample.

Eutrophication Enrichment of water by nutrients, especially compounds of nitrogen and phosphorus, which increases productivity of ecosystems, leading usually to lowering water quality and several adverse ecological and social effects (e.g., secondary pollution due to accelerated growth of algae and toxic cyanobcateria, depletion of oxygen).

Evapotranspiration Quantity of water transferred from the soil to the atmosphere by evaporation and plant transpiration.

Fecal coliform Bacteria present in the intestines or feces of warm-blooded animals. They often are used as indicators of the sanitary quality of the water. Their concentrations are expressed as number of colonies per 100 ml of sample.

Flocculation process Process by which clumps of solids in water or sewage aggregate through biological or chemical action so they can be separated from water or sewage.

Floodplain Nearly level land along a stream flooded only when the streamflow exceeds the water carrying capacity of the channel.

Floriculture The cultivation of flowers or flowering plants.

Geographic Information Systems (GIS) A computer-based system of principles, methods, instruments and geo-referenced data used to capture, store, extract, measure, transform, analyze and map phenomena and processes in a given geographic area.

Geomorphology The branch of geology dealing with the origin, evolution, and configuration of the natural features of the Earth's surface or a particular region of it.

Groundwater mining When discharge from an aquifer -usually due to groundwater pumping for municipal and business use- exceeds recharge.

Helminth Parasitic worms of the phylum platyhelminthes including digenetic flukes (class Trematoda, e.g., *Schistosoma* spp.) and tapeworms (class Cestoidea, e.g., *Taenia solium*, pork tapeworm).

Hydraulic conductivity Property of a saturated porous medium which determines the relationship, called Darcy's law, between the specific discharge and the hydraulic gradient causing it.

Hydrograph Graph showing the variation in time of some hydrological data such as stage, discharge, velocity, sediment load, etc. (hydrograph are mostly used for stage or discharge)

Hydrologic balance Inflow to, outflow from, and storage in, a hydrologic unit, such as a drainage basin, aquifer, soil zone, lake, reservoir, or irrigation project. (USGS)

Hydrological cycle Succession of stages through which water passes from the atmosphere to the earth and returns to the atmosphere: evaporation from the land or sea or inland water, condensation to form clouds, precipitation, accumulation in the soil or in bodies of water, and re-evaporation.

Impervious Having a texture that does not permit water to move through it perceptibly under static pressure ordinarily found in subsurface water.

Infiltration Flow of water through the soil surface into a porous medium, such as the soil, or from the soil into a drainage pipe.

Inorganic pollutant Mineral-based compounds such as metals, nitrates, and asbestos; naturally occurring in some water, but can also enter water through human activities.

Ion-exchange A common water-softening method often found on a large scale at water purification plants that remove some organics and radium by adding calcium oxide or calcium hydroxide to increase the pH to a level where the metals will precipitate out.

Isohyet Line joining the points where the amount of precipitation, in a given period, is the same.

Land degradation A human-induced or natural process which negatively affects the land to function effectively within an ecosystem, by accepting, storing and recycling water, energy, and nutrients.

Leacheate Water that collects contaminants as it trickles through wastes, pesticides or fertilizers. Leaching may occur in farming areas, feedlots, and landfills, and may result in hazardous substances entering surface water, groundwater, or soil.

Mollisols Mineral soils that have a mollic epipedon overlying mineral material with a base saturation of 50 per cent or more when measured at neutral pH (pH = 7).

Nitrification The process whereby ammonia in wastewater is oxidized to nitrite and then to nitrate by bacterial or chemical reactions.

Orographic precipitation Precipitation caused by the ascent of moist air over orographic barriers.
Oxidation process Chemical process which can lead to the fixation of oxygen or the loss of hydrogen, or the loss of electrons; the opposite is reduction.
Percolation Flow of a liquid through an unsaturated porous medium, e.g., of water in soil, under the action of gravity.
Perennial flow Stream which flows continuously all through the year.
Pervious; *syn.* permeable; Having a texture that permits water to move through it perceptibly under static pressure ordinarily found in subsurface water.
Pluvial Of, or relating to rain; characterized by much rain.
Point source A stationary location or fixed facility from which pollutants are discharged; any single identifiable source of pollution; e.g., a pipe, ditch, ship, ore pit, factory smokestack.
Pollutants Any substance introduced into the environment that adversely affects the usefulness of a resource or the health of humans, animals, or ecosystems.
Potabilization Process through which water that is made safe for drinking and cooking.
Protozoa One-celled animals that are larger and more complex than bacteria. May cause disease.
Rainwater harvesting Collection and concentration of rainwater to be used for irrigation of annual crops, pastures and trees; domestic consumption and livestock consumption.
Raw water Water which has received no treatment whatsoever, or water entering a plant for further treatment.
Reverse osmosis A treatment process used in water systems by adding pressure to force water through a semi-permeable membrane. Reverse osmosis removes most drinking-water contaminants. Also used in wastewater treatment.
Riparian Pertaining to the banks of a stream.
Rotaviruses Any member of the genus Rotavirus of wheel-shaped reoviruses which are pathogens of a wide range of mammals and birds, typically causing severe diarrhoeal illness, esp. in the young.
Runoff That part of precipitation that appears as streamflow.
Safe yield Amount of water (in general, the long-term average amount) which can be withdrawn from a groundwater basin or surface water system without causing undesirable results.
Salinity Measure of the concentration of dissolved salts, mainly sodium chloride, in saline water and sea water.
Saturated zone Part of the water-bearing material in which all voids, large and small, are filled with water.
Septic system An on-site system designed to treat and dispose of domestic sewage. A typical septic system consists of tank that receives waste from a residence or business and a system of tile lines or a pit for disposal of the liquid effluent (sludge) that remains after decomposition of the solids by bacteria in the tank and must be pumped out periodically.
Sewage; *syn,* wastewater; The waste and wastewater produced by residential and commercial sources and discharged into sewers.
Sewerage The entire system of sewage collection, treatment, and disposal.

Sheetflow Flow in a relatively thin sheet, of nearly uniform thickness, over the soil surface.
Siltation Process of filling up or raising the bed of a watercourse or body of water through deposition of sediments.
Sludge Accumulated solids separated from various types of water as a result of natural or artificial processes.
Soil subsidence Lowering in elevation of a considerable area of land surface, due to the removal of liquid or solid underlying material or removal of soluble material by means of water.
Stakeholder Any organization, governmental entity, or individual that has a stake in or may be impacted by a given approach to environmental regulation, pollution prevention, energy conservation, etc.
Stormwater Runoff from buildings and land surfaces resulting from storm precipitation.
Streamflow General term for water flowing in a stream or river channel.
Tectonically active Refers to ongoing rock-deforming processes and resulting structures that occur over large sections of the lithosphere.
Total Suspended Solids (TSS) In a water sample, the total weight of suspended constituents per unit volume or unit weight of the water.
Transmisivity rate Rate at which water is transferred through a unit width of an aquifer under a unit hydraulic gradient. It is expressed as the product of the hydraulic conductivity and the thickness of the saturated portion of an aquifer.
Unconfined aquifer Aquifer containing unconfined groundwater that is having a water table and an unsaturated zone.
Upconing Process by which saline water underlying freshwater in an aquifer rises upward into the freshwater zone as a result of pumping water from the freshwater zone.
Urban water cycle A water cycle including all the components of the natural water cycle with the addition of urban flows from water services, such as the provision of potable water and collection and treatment of wastewater and stormwater.
Vadose zone; *syn.* unsaturated zone; Subsurface zone above the water table in which the spaces between particles are filled with air and water, and the water pressure is less than atmospheric.
Vertisols Mineral soils that have 30 per cent or more clay, deep wide cracks when dry, and either gilgai microrelief, intersecting slickensides, or wedge-shaped structural aggregates tilted at an angle from the horizon.
Wastewater Water containing waste, i.e. liquid or solid matter discharged as useless from a manufacturing process.
Water cycle Succession of stages through which water passes from the atmosphere to the earth and returns to the atmosphere: evaporation from the land or sea or inland water, condensation to form clouds, precipitation, and accumulation in the soil or in bodies of water, and re-evaporation.
Water reclamation The restoration of wastewater to a state that will allow its beneficial reuse.
Water rights; *syn*: riparian rights; Entitlement of a land owner to certain uses of water on or bordering the property, including the right to prevent diversion or misuse of upstream waters.

Wetlands Lands where water saturation is the dominant factor determining the nature of soil development and the types of plant and animal communities living in the surrounding environment. Other common names for wetlands are bogs, ponds, estuaries, and marshes.

Xeric; *syn.* dry; An adjective often used to describe upland habitats with well-drained soils.

List of Contributors

Chapters 1–6
Larry W. Mays, Arizona State University, Tempe, Arizona, USA

Case Studies

- Blanca Jimenez, Universidad Nacional Autónoma de México, Mexico – Water and wastewater management in Mexico City: Challenges and options for integrated management
- Robert G. Arnold and Katherine P. Arnold, The University of Arizona, Tucson, Arizona, USA – Integrated urban water management in the Tucson, Arizona metropolitan area
- Messele Z. Ejeta[1], Getu F. Biftu[2] and Dagnachew A. Fanta[1], [1]California Department of Water Resources, Sacramento, California, USA; [2]Golder Associates Ltd., Alberta, Canada – Upper Awash River System, Ethiopia
- Jun Ma, Xiaohong Guan and Liqiu Zhang, Harbin Institute of Technology, Harbin, China – Water treatment for urban water management in China
- El Said M. Ahmed[1] and Mohamed A. Ashour[2], [1]Arizona Department of Water Resources, Phoenix, Arizona, USA; [2]Assiut University, Assiut, Egypt – Challenges for urban water management in Cairo, Egypt: The need for sustainable solutions

Chapter 1

Introduction

1.1 WATER SCARCITY IN ARID AND SEMI-ARID REGIONS

The purpose of this book is to examine the integrated management of water resources in urban environments in arid and semi-arid regions around the world. Integrated urban water resources management includes both water supply management and water excess management. The concept of water resources sustainability is used as an overall goal of integrated urban water resources management.

To begin with, we will identify the locations of arid and semi-arid regions in the world. Aridity is defined as a lack of moisture which is essentially a climatic phenomenon based upon the average climatic conditions over a region (Agnew and Anderson, 1992). Arid regions have been identified by climatological mapping. Of the many classifications based on climate, Meigs (1953) developed a set of maps for UNESCO that received wide international acceptance and were recognized by the World Meteorological Organization. The hyper-arid or extremely arid, the arid and the semi-arid regions of the world are shown in Figure 1.1.

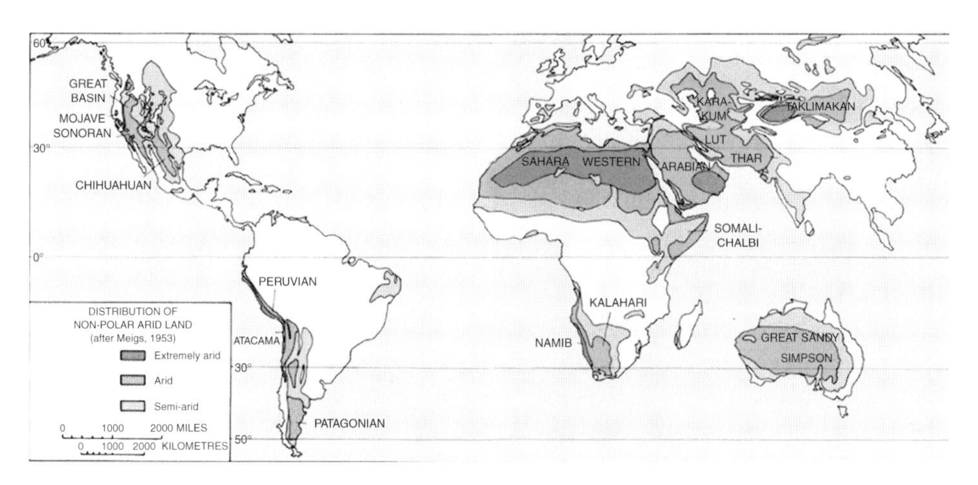

Figure 1.1 Distribution of non-polar arid land (after Meigs, 1953) (See also colour plate 1)

Source: http://pubs.usgs.gov/gip/deserts/what/world.html

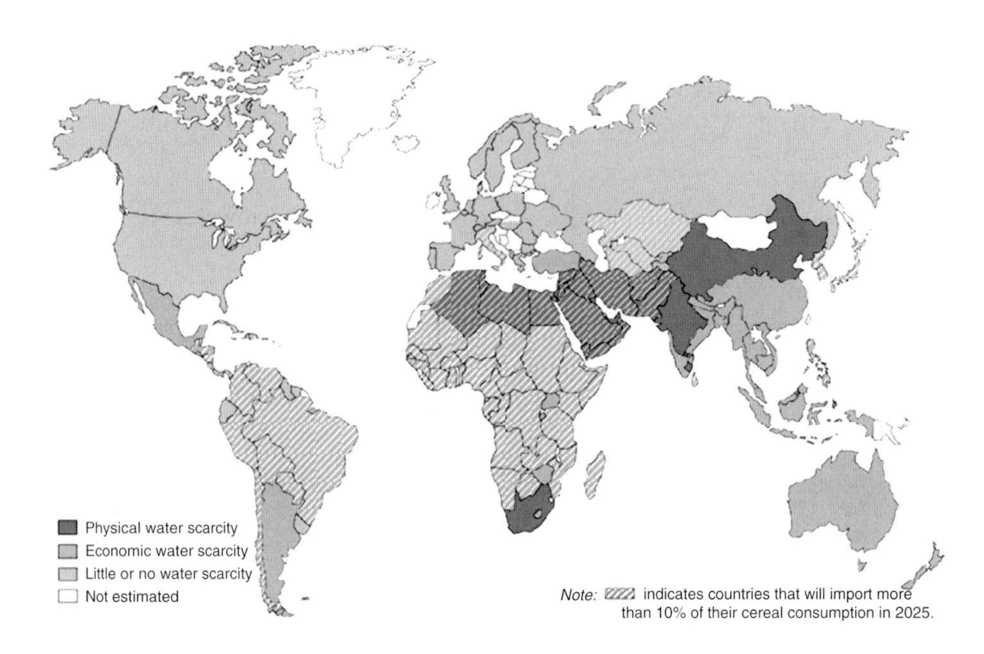

Figure 1.2 Projected water scarcity in 2025, International Water Management Institute (See also colour plate 2)

Source: http://www.iwmi.cgiar.org/assessment/files/pdf/publications/ResearchReports/CARR1.pdf

Urban populations are growing rapidly around the world, with the addition of many megacities (populations of 10 million or more inhabitants). In 1975 there were only four megacities in the world, whereas by 2015 there may be over 22 (Marshall, 2005). Other cities that will not become megacities are also growing very rapidly around the world. By 2010 more than 50% of the world's population is expected to live in urban areas (World Water Assessment Program, 2006). Many of the urban areas are in arid and semi-arid (ASA) regions of the world.

Urban populations demand high quantities of energy and raw material, water supply, removal of wastes, transportation, etc. Urbanization creates many challenges for the development and management of water supply systems and the management of water excess from storms and floodwaters. Many urban areas of the world have been experiencing water shortages, which are expected to explode this century unless serious measures are taken to reduce the scale of this problem (Mortada, 2005). Most developing countries have not acknowledged the extent of their water problems. This is evidenced by the absence of any long-term strategies for water management.

Water scarcity is certainly one of the major challenges in urban areas in arid and semi-arid regions of the world. Figure 1.2 shows regions with projected water scarcity around the world by 2025. Regions suffering from both physical and economic scarcity are illustrated in the figure. Balancing water-scarcity and population (human demand) is the major challenge in many arid and semi-arid regions of the world.

The Middle East and North Africa (MENA) region is the driest water-scarce region in the world. MENA is home to 6.3% of the world's population and contains only 1.4% of the world's renewable freshwater (Roudi-Fahima et al., 2002). To complicate matters, as the population pressures in the region increase, demand for water increases. How is this demand to be met given the scarce water resources of the region?

The MENA region is one of great contrast as some of the countries, mainly in Africa, are extremely poor whereas many of the countries in the Middle East have very strong financial institutions. Access to water and sanitation is variable in this region. Libya, Tunisia and the countries of the Gulf Cooperation Council have the highest rate of access to safe drinking water at over 90%. In comparison, the lowest access is found in Mauritania, Palestine, and Yemen (Stedman, 2006).

1.2 IN THE BEGINNING

Humans have spent most of their history as hunting and food-gathering beings. Only in the last 9,000 to 10,000 years have we discovered how to raise crops and tame animals. This revolution probably first took place in the hills to the north of present day Iraq and Syria. From there the agricultural revolution spread to the Nile and Indus Valleys. During this time of agricultural revolution, permanent villages took the place of a wandering existence. About 6,000 to 7,000 years ago, farming villages of the Near East and Middle East became cities. The first successful efforts to control the flow of water were made in Mesopotamia and Egypt. Remains of these prehistoric irrigation canals still exist. About 5,000 years ago, the science of astronomy was born and observations of other natural phenomena led to knowledge about water, resulting in advances for its control and use.

In ancient Mesopotamia, every city of the Sumer and Akkud (4th millennium BC) had a canal(s) connected to the Euphrates River or a major stream for both navigation and water supply for daily uses. In Mari a canal was connected to the city from both ends and passed through the city (Viollet, 2006). Servant women filled the 25 m^3 cistern of the palace with water supplied by the canal. Later on, other cisterns were built in Mari and connected to an extended rainfall collection system. Terracotta pipes were used in Habuba Kebira (in modern Turkey), a Sumerian settlement in the middle of the Euphrates valley in the middle of the 4th millennium BC (Viollet, 2006). In the 3rd millennium BC, the Indus civilization had bathrooms in houses and sewers in streets. The Mesopotamians were not far behind (Adams, 1981). In the 2nd millennium BC, the Minoan civilization on Crete had running water and flushing latrines (Evans, 1964). The Minoan and Mycenaean settlements used cisterns 1,000 years before the classical and Hellenistic-Greek cities. Cisterns were used to supply (store runoff from rooftops) water for the households through the dry summers of the Mediterranean. Brief histories of ancient water distribution are given in Mays (2006) and Mays et al. (2007).

The development of groundwater dates from ancient times. Other than dug wells, groundwater in ancient times was supplied from horizontal wells known as *qanats*, also known as a *kanerjing* (western China), *karez* (Afghanistan and Pakistan), *khittara* (Morocco), *fogara* (Arabia), and *falaj* (Northern Africa). The *qanat* spread from Persia to other locations over many years. Typically, a gently sloping tunnel through alluvial material leads water by gravity flow from beneath the water table at its upper end to a ground surface outlet at its lower end. Vertical shafts dug at closely spaced intervals provide access to the tunnel. *Qanats* are laboriously hand constructed, employing techniques that date back over 3,000 years. To illustrate the tremendous effort involved in constructing a *qanat*, Beaumont (1971) estimated the volume of material excavated for the 29 km *qanat* near Zarand, Iran as 75,400 m^3. This *qanat* had a mother (furthest upstream) well with a depth of 96 m and 966 shafts along its length.

For thousands of years the people of Egypt have owed their very existence to a river that flowed mysteriously and inexplicably out of the greatest and most forbidding desert in the world (Hillel, 1994). Herodotus said that 'Egypt is an acquired country, the gift of the River'. The ancient Egyptians not only depended on the Nile for their livelihoods, but they also considered the Nile to be a deific force of the universe, to be respected and honoured if they wanted it to treat them favourably. Its annual rise and fall were likened to the rise and fall of the sun, each cycle equally important to their lives, though both remaining a mystery. The first actual recorded evidence of water management was the illustrated mace head of King Scorpion, the last of the Pre-dynastic kings, which has been interpreted as depicting a ceremonial start to breaching the first dyke to allow water to inundate the fields or the ceremonial opening of a new canal (Strouhal, 1992). Similarly, others have interpreted the main part of the mace head of the king as depicting irrigation work under his supervision. This mace head indicates that the ancient Egyptians began practising some form of water management for agriculture about 5,000 years ago. Throughout history there were advancements in the irrigation of the Nile, from natural irrigation to artificial irrigation to the development of lift irrigation with the *shaduf* and then the Archimedes screw (or *tanbur*) and the *saqiya* (or waterwheel). From a water management perspective, all evidence known suggests that flood control and irrigation, at the social and administrative levels, were managed locally by the rural population within a basin.

The rise and sustainability of Egypt, with so many great achievements, was based primarily on the cultivating of grain on the Nile River floodplain, without a centralized management of irrigation. What is unique is that Egypt probably survived for so long because production did not depend on a centralized state. Collapses of the government and changes of dynasties did not undermine irrigation and agricultural production on the local level. 'The secret of Egyptian civilization was that it never lost sight of the past' (Hassan, 1998).

Urban hydraulic systems started to develop in the Bronze Age and particularly in the mid-third millennium BC in an area extending from India to Egypt. About the same time, advanced urban water technologies developed in Greece and particularly on the island of Crete where the Minoan civilization was flourishing. These included the construction and use of aqueducts, cisterns, wells, fountains, bathrooms and other sanitary facilities, which suggest lifestyle standards close to those of the present day. The technology matured and evolved during the latter stages of Greek civilization with a peak during the Hellenistic period, supported by the understanding of natural processes and the development of scientific concepts. The Romans developed sophisticated engineering skills and were able to expand these technologies on large-scale projects throughout their large Empire.

After the fall of the Roman Empire, the concepts of science and technology related to water resources probably retrogressed. Water supply systems, water sanitation and public health declined in Europe, which entered a period known as the Dark Ages. Historical accounts tell of extremely unsanitary conditions – polluted water, human and animal wastes in the streets, and wastewater thrown out of windows onto passers-by. Various epidemics ravaged Europe. During the same period, several Byzantine sites in Greece and Asia Minor maintained a high level of civilization and Islamic cultures, on the periphery of Europe, had religiously mandated high levels of personal hygiene, along with highly developed water supply, sewerage and adequate sanitation systems. Europe regained high standards of water supply and sanitation only in the nineteenth century.

1.3 THE URBAN WATER CYCLE AND URBANIZATION

Urbanization is a reality of our changing world. From a water resources perspective, urbanization causes many changes to the hydrological cycle, including radiation flux, amount of precipitation, amount of evaporation, amount of infiltration, and increased runoff. The changes in the rainfall-runoff components of the hydrologic cycle can be summarized as follows (Marsalek et al., 2006):

- transformation of undeveloped land into urban land (including transportation corridors
- increased energy release (i.e. greenhouse gases, waste heat, heated surface runoff)
- increased demand on water supply (municipal and industrial).

The overall urban water cycle is illustrated in Figure 1.3, showing the main components and pathways. How does the urbanization process change the water budget from predevelopment to developed conditions of the urban water cycle in arid and semi-arid regions? This change is a very complex process and difficult to explain.

The process of urbanization often causes changes in groundwater levels because of a decrease in recharge and increased withdrawal. Three major conditions disrupt the

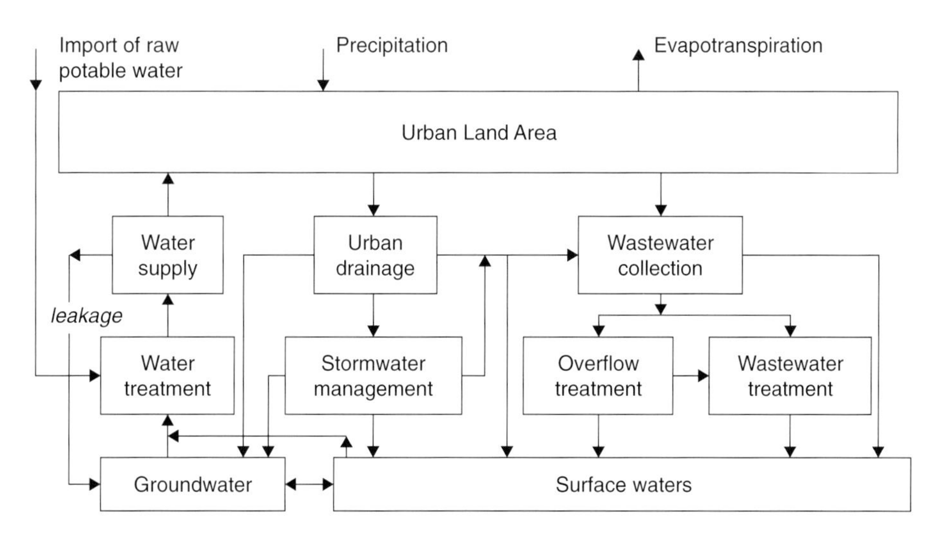

Figure 1.3 Urban water cycle – Main components and pathways

Source: Urban Water Cycle: Processes and Interactions, by Marsalek et al., IHP-VI, Technical Publications in Hydrology, No. 78, UNESCO, Paris, 2006

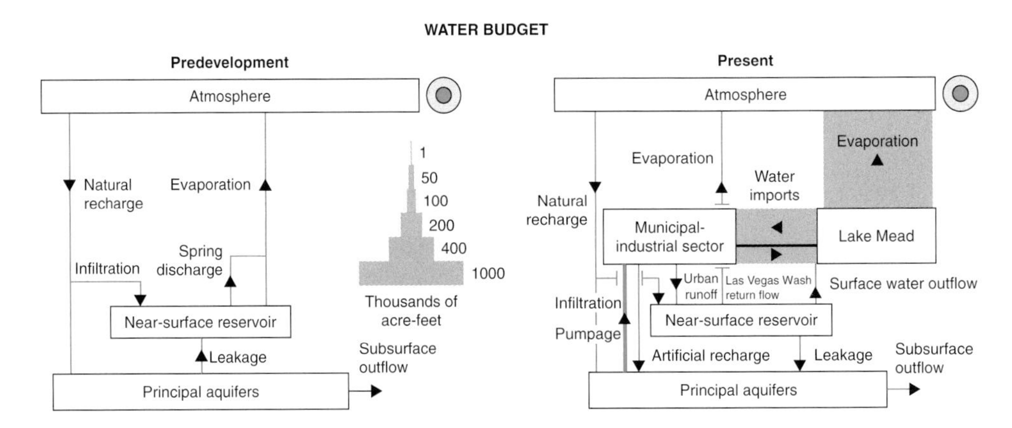

Figure 1.4 Water budget for Las Vegas, Nevada comparing the predevelopment and present conditions of urbanization.

Source: Pavelko et al., 1999; Courtesy of the US Geological Survey

subsurface hydrologic balance and produce declines in groundwater levels (Todd and Mays, 2005):

- reduced groundwater recharge due to paved surface areas and storm sewers
- increased groundwater discharge by pumping wells
- decreased groundwater recharge due to export of wastewater collected by sanitary sewers.

Decreased groundwater levels can cause land subsidence, such as the severe land subsidence described in the Mexico City case study. Groundwater pollution is another consequence of urbanization, which has been experienced in many arid and semi-arid regions of the world. Groundwater pumping in many coastal areas in arid and semi-arid regions has caused severe saltwater intrusion, making the water unusable for municipal water supplies.

In many locations in arid and semi-arid basins around the world, the development of groundwater resources to support urban growth and land use changes have drastically altered the way water circulates through the basin. One such location is Las Vegas, Nevada. Figure 1.4 compares the predevelopment and present water budgets for Las Vegas, Nevada. The present water budget reveals that only a small fraction of the water used in Las Vegas Valley is actually consumed, and therefore removed from the water cycle, by domestic, agricultural and municipal/industrial uses. Most of the water is either returned to the aquifer system, evaporated, or discharged into the Colorado River system of generally poorer quality. Large quantities of this generally poorer-quality water drain from over-watered lawns, public sewers, paved surfaces and drainage ways. The urban runoff flows onto open ground where it evaporates, is transpired by plants, or recharges the near-surface reservoir. Large amounts of treated sewage water are discharged into the Colorado River system. Groundwater has been depleted in the principal aquifers and aquitards, causing land subsidence, while the shallow, near-surface reservoir has been recharged with poor-quality urban runoff (Pavelko et al., 1999).

1.4 THE INTEGRATED URBAN WATER SYSTEM

'Urban water system' implies that there is a single urban water system that is an integrated whole. The concept of a single 'urban water system' is not fully accepted because of the lack of integration of the various components that make up the total urban water system. For example, in municipalities it is common to plan, manage and operate urban water as separate entities, such as by service, i.e. water supply, wastewater, flood control and stormwater. Typically there are separate water organizations and management practices within a municipality, or local or regional government, because that is the way they have been historically. Grigg (1986) points out that integration could be by functional integration and area-wide integration. There are many linkages between the various components of the urban water system, the hydrologic cycle being what connects the urban water system together. There are many reasons why the urban water system should be considered in an integrated manner. Two of the principal reasons are the natural connectivity of the system through the hydrologic cycle and the real benefits that are realized through integrated management rather than independent action.

The urban water management system is considered herein as two integrated major entities, water supply management and water excess management. The various interacting components of water excess and water supply management in conventional urban water infrastructure are:

Water supply management

- Sources (groundwater, surface water, reuse)
- Transmission
- Water treatment (WT)
- Distribution system
- Wastewater collection
- Wastewater treatment (WWT)
- Reuse

Water excess management

- Collection/drainage systems
- Storage/treatment
- Flood control components (levees, dams, diversions, channels).

Balancing water scarcity and human demand in many areas may require the use of both old and new technologies. The book, *The Water Atlas: Traditional Knowledge to Combat Desertification*, by Pietro Laureano (2005), addresses the use of old technologies (traditional methods) in the meeting of water demands. In the Middle East and North Africa, these traditional methods include *qanats*, rainwater harvesting and other technologies. Newer technologies include desalination, treatment and reuse of wastewater, reallocating water from agriculture to domestic and industrial sectors, using less water-intensive crops, managing water demand, conservation, increasing distribution efficiencies, instituting legal and institutional reforms, slowing population growth and others.

1.5 INTEGRATED URBAN WATER MANAGEMENT (IUWM): THE BIG PICTURE

1.5.1 Definition of IUWM

Integrated urban water management (IUWM) is 'a participatory planning and implementation process, based on sound science, which brings together stakeholders to decide how to meet society's long-term needs for water and coastal resources while maintaining essential ecological services and economic benefits' (USAID Water Team, http://www.gdrc.org/uem/water/iwrm/index.html).

According to the USAID Water Team, the principal components of an integrated urban water resources system include:

- *Supply optimization*, including assessments of surface and groundwater supplies, water balances, wastewater reuse and environmental impacts of distribution and use options.
- *Demand management*, including cost-recovery policies, water use efficiency technologies and decentralized water management authority.
- *Equitable access* to water resources through participatory and transparent management, including support for effective water users associations, involvement of marginalized groups and consideration of gender issues.
- *Improved policy, regulatory and institutional frameworks*, such as the implementation of the polluter-pays principle, water quality norms and standards and market-based regulatory mechanisms.
- *Intersectoral approach* to decision making, combining authority with responsibility for managing the water resource.

The following excerpt on integrated water resources management (IWRM) is from the *Industry Sector Report for the World Summit on Sustainable Development (WSSD)* prepared by the International Water Association (IWA) (http://www.gdrc.org/uem/water/iwrm/index.html).

> The fundamental premise is generally accepted that IWRM should be applied at catchment level, recognising the catchment or watershed as the basic hydrological unit of analysis and management. At implementation level, there is a growing conviction that integrated urban water management (IUWM) could be pursued as a vital component of IWRM within the specific problematic context of urban areas.
>
> Cities are dominant features in the catchments where they occur, and successes in IUWM will make important contributions to the theory and practice of integrated catchment management (ICM) and IWRM in the broader basin context. Thus, IUWM is not seen as a goal in itself, but rather a practical means to facilitate one important subsystem of the hydrological basin. IUWM must inter alia endeavour to optimise the interfacing of urban water concerns with relevant activities beyond the urban boundaries, such as rural water supply, downstream use, and agriculture.
>
> IUWM means that in the planning and operation of urban water management, consideration should be given to the interaction and collective impact of all

water-related urban processes on issues such as human health; environmental protection; quality of receiving waters; water demand; affordability; land and water-based recreation; and stakeholder satisfaction. In addition, IUWM requires involvement by stakeholders such as those responsible for water supply and sanitation services, stormwater and solid waste management, regulating authorities, householders, industrialists, labour unions, environmentalists, downstream users, and recreation groups. While local authorities are well placed to initiate and oversee IWRM/IUWM programmes, planning and implementation should be driven by a combination of top-down regulatory responsibility and bottom-up user needs/obligations. Top-heavy governmental approaches are to be discouraged because they become bureaucratic and unresponsive to the concerns of water users.

1.5.2 A word on integration

Integration has many meanings in the water management context (Cabrera and Lund, 2002). Integrated water management involves many aspects of integration: *sources* (multiple water sources considered?); *variability of sources* (planning for the entire range of wet to dry conditions?); *supply and demand aspects* (water supply augmentation, water supply conservation/demand management, etc considered?); *sources and sinks* (wastewater as a source, reuse potential, etc); *scale* (local, city, regional user levels jointly considered?); *responsibility, coordination and implementation* (water users, water supply, water treatment and delivery, wastewater collection and treatment, stormwater management, flood control management considered together by one entity?). This list is only partially complete as many other aspects are involved.

Ideally, integrated urban water management would be integrated across all these aspects although it is usually impossible to have a completely integrated urban water management. However, there remains considerable potential for improving urban water management through a better 'integrated' consideration of the various aspects and options. A truly integrated urban water management is practically impossible, whether it is in a developed region such as the south-western United States or in developing regions in MENA (Middle East and North Africa).

1.5.3 An example of water-related challenges in MENA

The key water-related challenges in the MENA (Middle East and North Africa) region include (http://wbln0018.worldbank.org/mna/mena.nsf; Stedman, 2006):

Water resources planning and management

— Resource: supply augmentation to manage scarcity, variability and quality.
— Demand: improving efficiency and cost recovery in complex and politically sensitive atmospheres.
— Allocation: among irrigated agriculture, urban and environment uses. (A complex challenge requiring that rights, trade, incentives and economics be addressed.)

Water and institutions

— Policy: legal, institutional and regulatory frameworks to manage resources productively require reform.
— Sustainability: social and environmental issues need to be better addressed.

Water and finance

— Financing, operation and maintenance: deals with tariffs, cost sharing/recovery, private sector participation and national budget.
— High cost of next-best options for supply augmentation (e.g., treated wastewater reuse, inter-basin water transfer, desalination of sea water) with poor prospects of financing.

1.5.4 The Dublin principles

The following four simple, yet powerful messages, were provided in 1992 in Dublin. They were the basis for the Rio Agenda 21 and for the millennium Vision-to-Action:

1. *Freshwater is a finite and vulnerable resource, essential to sustain life, development and the environment*, i.e. one resource, to be holistically managed.
2. *Water development and management should be based on a participatory approach, involving users, planners, and policymakers at all levels*, i.e. manage water with people – and close to people.
3. *Women play a central role in the provision, management and safeguarding of water*, i.e. involve women all the way!
4. *Water has an economic value in all its competing uses and should be recognized as an economic good*, i.e. having ensured basic human needs, allocate water to its highest value, and move towards full cost pricing, rational use and cost recovery.

Poor water management hurts the poor most! The Dublin principles aim at wise management with a focus on poverty.

1.6 WATER RESOURCES SUSTAINABILITY

The overall goal of integrated urban water resources management should be water resources sustainability. In, *Water Resources Sustainability*, Mays (2007) defines water resources sustainability as follows:

> Water resources sustainability is the ability to use water in sufficient quantities and quality from the local to the global scale to meet the needs of humans and ecosystems for the present and the future to sustain life, and to protect humans from the damages brought about by natural and human-caused disasters that affect sustaining life.

Because water affects so many aspects of our existence, whichever definition is used, there are many facets that must be considered. These are summarized as follows:

- Water resources sustainability includes the availability of freshwater supplies throughout periods of climatic change, extended droughts, population growth, while leaving the needed supplies for the future generations.
- Water resources sustainability includes having the infrastructure to provide a water supply for human consumption and food security, and providing protection from water excess such as floods and other natural disasters.

- Water resources sustainability includes having the infrastructure for clean water and for treating water after it has been used by humans before being returned to water bodies.
- Water sustainability requires adequate institutions for both the water supply management and water excess management.
- Water sustainability can be defined on a local, regional, national or international basis.

Sustainable water use has been defined by Gleick et al. (1995) as 'the use of water that supports the ability of human society to endure and flourish into the indefinite future without undermining the integrity of the hydrological cycle or the ecological systems that depend on it'. The following seven sustainability requirements were presented:

- A basic water requirement will be guaranteed to all humans to maintain human health.
- A basic water requirement will be guaranteed to restore and maintain the health of ecosystems.
- Water quality will be maintained to meet certain minimum standards. These standards will vary depending on location and how the water is to be used.
- Human actions will not impair the long-term renewability of freshwater stocks and flows.
- Data on water-resources availability, use and quality will be collected and made accessible to all parties.
- Institutional mechanisms will be set up to prevent and resolve conflicts over water.
- Water planning and decision making will be democratic, ensuring representation of all affected parties and fostering direct participation of affected interests.

Howard (2002) feels that water resources sustainability must be considered within the framework of probability, and that reliability in itself is an inadequate measure of sustainability. He states that, 'risk provides a more comprehensive definition of what encompasses both reliability and costs of shortages'. He defines a sustainable system as one that maintains acceptable risks over an indefinite time horizon. Sustainability as measured by risks has three main components for water management: probability of water supply shortages; costs when shortages are encountered; and level of acceptability of the risks.

Sustainable urban water systems are being advocated because of the depletion and degradation of urban water resources coupled with the rapid increases in urban populations around the world. Marsalek et al. (2006) defined the following basic goals for sustainable urban water systems:

- Supply of safe and good-tasting drinking water to the inhabitants at all times.
- Collection and treatment of wastewater in order to protect the inhabitants from diseases and the environment from harmful impacts.
- Control, collection, transport and quality enhancement of stormwater in order to protect the environment and urban areas from flooding and pollution.
- Reclamation, reuse and recycling of water and nutrients for use in agriculture or households in case of water scarcity.

In North America and Europe, most of the goals of sustainability have been achieved or are within reach. In developing parts of the world, the goals of sustainability are far from being achieved. As mentioned previously the Millennium Development Goals put a strong emphasis on poverty reduction and reduced child mortality.

1.7 FOCUS OF CASE STUDIES

Case studies are presented in this book in order to emphasize the various aspects of urban water management in arid and semi-arid locations around the world. These case studies include: Mexico City, Mexico; Tucson, Arizona Metropolitan Area; Awash River Basin, Ethiopia; China and Cairo, Egypt.

1.7.1 Case study 1: Water and wastewater management in Mexico City: Challenges and options for integrated management

This case study by Dr. Blanca Jimenez focuses on water and wastewater management in Mexico City, Mexico, a megacity experiencing many challenges in urban water management. Mexico City is unique in the range of challenges faced:

- Water transfers from other basins, causing problems in the basins from which water was transferred.
- Over-exploitation of groundwater, causing huge soil subsidence and resulting in loss of sewage and drainage capacity, leaks in water distribution and wastewater collection networks, deterioration of groundwater quality and other problems such as serious structural problems with buildings and the re-levelling of metro rails.
- Very large leakage from the water distribution system (37–40% of water conveyed is lost to leakage, 23 m^3/s loss).
- Of the wastewater produced (67.7 m^3/s) only 11% (7.7 m^3/s) is treated. Untreated wastewater is transferred to the Tula Valley 100 km north of Mexico City for irrigation purposes (agricultural use), resulting in serious health problems.
- Because of the large recharge of untreated wastewater (13 times the natural recharge) in the Tula Valley, the water table has risen and the Tula River flow increased significantly. Water quality of the recharged wastewater is reasonably safe; however, chlorination is not the best option for disinfection.
- Wastewater recharge has formed a new watercourse and completely modified the ecology of the Tula Valley from a semi-desert area to one with springs and wetlands.

Options for integrated water management would include the development of a new metropolitan water authority with the following activities: controlling of soil subsidence; protection of groundwater quality; the reduction of large leakage in the water distribution system; implementation of aggressive and innovative wastewater reuse programmes; innovative and comprehensive educational programmes; improvement of economic tools; rainwater harvesting; and implementation of professional public participation programmes.

1.7.2 Case study II: Integrated urban water management in the Tucson, Arizona Metropolitan Area

This case study, by Dr. Robert Arnold and Ms Katie Arnold, focuses on Tucson, Arizona, a very fast growing city with a unique water infrastructure and water management arena. Prior to 2001, Tucson relied entirely on groundwater to meet potable water demands. Because of the high quality of the groundwater, only disinfection was required for treatment. The over-reliance on groundwater was causing the water table to decline significantly. In Arizona the Groundwater Management Act (GMA) mandates the attainment of 'safe yield', which is a balance between groundwater withdrawals and replenishment, by 2025. Water demand has outgrown the renewable groundwater supply. The delivery of Colorado River water by the Central Arizona Project (CAP) is the only possibility for 'sustained growth'. However, Arizona has the lowest priority use of CAP water making it vulnerable to water shortages under drought conditions.

The CAP water is recharged and recovered before it is used by the public. The underground storage of CAP water includes the Central Avra Valley Storage and Recovery Project, consisting of recharge basins, recovery wells, a booster station and a reservoir and associated piping. A new storage and recovery project is being planned. A mothballed water treatment facility could be revitalized for surface treatment of CAP water, accelerating the use of the water.

1.7.3 Case study III – Upper Awash River System, Ethiopia

Case study III, by Drs. Messele Z. Ejeta, Getu F. Biftu and Dagnachew A. Fanta, assesses the water resources of the Upper Awash River system in Ethiopia, providing perspectives for integrated water resources management. Water resources of this region are affected by long dry spells, and high intensity and short duration rainfall, which are important factors in the efficient operation of reservoirs for both urban and agricultural water allocation. The seasonality and exceptional spatial and temporal variation of the rainfall, combined with a limited infrastructure for storage and poorly protected watersheds, expose millions of people to the threats of droughts and floods. There has not been an adequate exploration of the groundwater resources of this region; however, rich salt deposits and active volcanoes make the groundwater unsuitable for municipal and irrigation purposes.

Municipalities in the Upper Awash River region include Adama, Wanji and Matahara, which all get their water from the Awash River. Adama, one of the largest cities in Ethiopia, obtains its water from below the Koko dam. Because of the net rural to urban population migration, the urban water demand is significantly increasing. The Awash Basin Water Resources Administration Agency coordinates, administers, allocates and regulates the use of surface water resources of the basin. The practical significance of this young (1998) agency, with its unclear regulating requirements, is yet to be observed. To complicate matters further, the water laws of Ethiopia are not well understood.

The freshwater resources of the region are burdened with both point source and distributed source pollutants due to the fluvial soil characteristics, the disproportionate concentration of the country's industry in the basin and the active volcano in the Middle Awash. Evapotranspiration exceeds the mean annual rainfall leading to the accumulation of groundwater salts on the surface.

Ethiopian national drinking-water quality standards follow the World Health Organization (WHO) guidelines. Adama's water supply from the Awash River is treated using sedimentation basins, sand filters, and biological and chemical treatment facilities. Water is stored in service reservoirs and then delivered to the distribution system by gravity. The smaller communities such as Wanji use untreated groundwater for domestic purposes.

Only 13% of the people in Ethiopia have access to sanitation, 10% in the rural areas and nearly 45% in the urban areas, so that only about 1 in 7 people has access to improved sanitation. In the Upper Awash River region, sewage collection, treatment and disposal are a very crucial problem for the cities. Wastewater treatment from public sewers is almost nonexistent. Diarrhoea and parasitic infection continue to be a problem. Pit latrines are used commonly for household wastewater disposal facilities.

In Adama there are no stormwater protection facilities, other than storm drainage gutters, which are commonly used as disposal outlets for household wastewater. Adama lies at the foot of escarpments from which runoff flows to the city causing significant flooding.

In summary, water shortages are a result of the seasonality of the rainfall and the lack of infrastructure to store excess runoff during the flood season. Studies to access the sustainability of the Upper Awash River System are needed. The institutional capacity for integrated water resources planning and management in the Upper Awash River region is emerging. The impacts of climate change on the water resources will make management of the Upper Awash River Basin even more difficult.

1.7.4 Case study IV – Water treatment for urban water management in China

Case study IV, by Drs. Jun Ma, Xiaohong Guan and Liqiu Zhang, focuses on urban water management in the arid and semi-arid regions of China. Northern China has only 20% of the total water resources of China but supports more than half the total population. Over 80% of China's water resources are concentrated in the south-eastern part of the country, where the water resources are polluted and eutrophicated.

In 1986, 181 cities in China suffered water shortages and in 1995, 333 (about 50% of Chinese cities) suffered water shortages. Presently over 400 of the 660 cities in China have a water shortage, of which a large proportion results from water quality deterioration from pollution.

Conventional water treatment processes are used when water quality is good enough and modified conventional or advanced treatment processes are used for polluted water sources. Surface water is the main source of urban water supply in the arid and semi-arid regions of China, using conventional water treatment processes. In certain cases when the raw water is polluted, enhanced treatment processes such as preoxidation with permanganate, adsorption with powdered activated carbon, dissolved air flotation, biological filtration, etc. are used.

Wastewater treatment is a different story in that 297 cities, and more than 50,000 towns and villages, still have no wastewater treatment plants. By June 2005 there were only 708 wastewater treatment plants in 661 cities with a total capacity of only 49.12 million m^3/day.

Both combined and separate sewer systems are used, with most of the cities in the arid and semi-arid region using combined sewer systems. These combined systems have resulted in high pollution and hygienic risks, so that newly constructed systems are generally separate sewer systems. Even the separate systems cause adverse effects such as sources of pollution in lakes and rivers at receiving waters.

Rapid urbanization has compounded the problem, causing water scarcities and drastic conflicts between water supply and demand, and making water a key limiting factor of the urbanization process, as well as socio-economic development. With the huge production of sewage, municipal wastewater treatment has drawn more attention. The pollution of China's rivers is severe, and overall environmental quality is deteriorating fast. The water resources shortage has become a limiting factor for further development in China.

1.7.5 Case study V: Challenges for urban water management in Cairo, Egypt: The need for sustainable solutions

Case Study V, by Drs. El Said M. Ahmed and Mohamad A. Ashour, focuses on Cairo, Egypt which has the Nile River as its source of water. Since the 1970s and 1980s the water quality of the Nile has deteriorated due to increased industrial and agricultural discharges and some contamination from human sewage. The Cairo Water Authority has 13 water treatment plants distributed throughout the city. A majority of Cairo residents receive treated drinking water through individual connections in their homes. Many apartment buildings do not have distribution of water to individual apartments. People must wait in long lines at communal taps, or where there are no communal taps poorer people must buy water through unsafe containers at very high prices.

Rapid expansion of the city has placed a large strain on the water supply which has led to low pressure. Erratic pressure variations lead to contamination entering the water distribution and storage system from contaminated groundwater or sewage from leaking drains and sewers. Contamination enters the distribution system through damaged joints and cracks in the pipes. Many of the water distribution pipes are located adjacent to pipes carrying raw sewage and leaks are a constant problem with the aging infrastructure.

The wastewater system consists of six wastewater treatment plants. Wastewater in many areas is still conveyed through agricultural drains, thus shifting environmental problems to other regions. Daily water usage far exceeds the capacity of the sewage system, leading to standing pools of raw sewage in the streets and in the underground water table, as well as leaks to the Nile River and other sources of clean water.

Stormwater management systems are almost nonexistent. One approach to the future has been the establishment of 'new cities' such as New Cairo, which has a large growth area with two to three million people expected to live there.

Chapter 2

Arid and semi-arid regions: what makes them different?

2.1 PHYSICAL FEATURES

2.1.1 What is aridity?

Arid means dry, or parched and the primary determinant of aridity in most areas is the lack of rainfall (Slatyer and Mabbutt, 1964). Meigs (1953) divided xeric environments into extremely arid, arid and semi-arid (see Figure 1.1). Arid areas were defined as those in which the rainfall is not adequate for regular crop production; and semi-arid areas as those in which the rainfall is sufficient for short-season crops and where grass is an important element of the natural vegetation. To avoid confusion, the term desert is based upon land surface characteristics and can be considered as areas of low or absent vegetation cover with an exposed ground surface (Goudie, 1985). Agnew and Anderson (1992) considered the arid realm to encompass arid and semi-arid environments from desert through to steppe landscapes. They used the term desert to convey hyper-arid conditions where rainfalls are particularly low and vegetation is sparse.

2.1.2 Geomorphology

From a geomorphological viewpoint, no single process dominates arid environments. Arid lands vary from tectonically active mountainous regions in North and South America to the geologically stable shield areas in Africa and Australia. Dick-Peddie (1991) presented summary discussions of each continent along with the extent and vegetation of the semi-arid, arid, hyper-arid and riparian habitats within each region. The boundaries of these regions conform closely to Meigs (1953) arid, hyper and homoclimates.

Agnew and Anderson (1992) defined the following features of arid zone landscapes based upon Goudie (1985), Heathcote (1983) and Thomas (1989):

- Alluvial fans are fan-shaped deposits found at the foot of the slope, grading from gravels and boulders at the apex to sand and silt at the foot, called a bajada when coalesced.
- Dunes are aeolian deposits of sand grains (unconsolidated mineral particles) forming various shapes and sizes, depending upon the supply and characteristics of the material and the wind system.

- Bedrock fields including pediment, a piano-concave erosion surface sloping from the foot of an upland area; and hamada, a bare rock surface with little or no vegetation or surficial material.
- Desert flats with slight slopes possibly containing sand dunes, termed playa when the surface is flat and periodically inundated by surface runoff.
- Desert mountains are the most common feature of arid lands.
- Badlands well dissected, unconsolidated or poorly cemented deposits with sparse vegetation.

2.1.3 Soil characteristics

Soil characteristics in arid and semi-arid regions are influenced primarily by low rainfalls, high evaporation rates and low amounts of vegetation. The soils, therefore, have low organic matter, an accumulation of salts at the surface, little development of clay minerals, a low cation-exchange capacity, a dark or reddish colour due to desert varnish, and little horizon development due to the lack of percolating water (Fuller, 1974). Even though there are vast areas covered by thin, infertile soils, there are arid lands where soils are highly productive having a very high potential for agriculture.

Dregne (1976) presents the following percentages of cover soils:

- Entisols cover 41.5% of arid lands (immature soils ranging from barren sands to very productive alluvial deposits.
- Aridisols cover 35.9% of arid lands (red-brown desert soils, dry and generally only suitable for grazing without irrigation)
- Vertisols cover 4.1% of arid lands (moderately deep swelling clay which is difficult to cultivate).

According to Agnew and Anderson (1992), mollisols cover 11.9% of arid lands and alfisols cover 6.6%. Mollisols are one of the world's most important agricultural soils. Alfisols have a high base saturation, reasonably high clay contents and are agriculturally productive.

Because of the climate in arid lands, soil formation is dominated by physical disintegration with only slight chemical weathering. Elgabaly (1980) defines three main types of soil found in arid lands:

- Saline soils – Characterized by the presence of excess neutral salts (pH less than 8.5) that accumulate on the surface as a loose crust, depending upon the depth and salinity of the groundwater table.
- Saline-alkaline soils – Characterized by the presence of excess soluble salts (pH approximately equal to 8.5). The structure is more compact at a certain depth and darker in colour.
- Sodic soils – Characterized by the presence of low soluble salts (pH greater than 8.5). Surface colour is usually darker and clay accumulates in the B-horizon and a columnar structure eventually develops.

The formation or origin of salt-affected soils is connected with:

- climate, as saline soils are an element of arid lands
- relief, as saline soils are more common in low lands such as deltas and floodplains

- geomorphology and hydrology, as saline soils are related to the depth of the water table.

2.2 CLIMATE

2.2.1 Causes of aridity

One of several processes can lead to aridity; however, Hills (1966) believes that the major cause of aridity is explained through the global atmospheric circulation patterns. Thompson (1975) lists four main processes that explain aridity as presented by Agnew and Anderson (1992):

- High pressure – Air that is heated at the equator rises, moves polewards and descends at the tropical latitudes around 20 to 30 degrees latitude. This descending air is compressed and warmed, thus leading to dry and stable atmospheric conditions covering large areas such as the Sahara Desert.
- Wind direction – Winds blowing over continental interiors have a reduced opportunity to absorb moisture and will be fairly stable with lower humidity. These typically dry, northeasterly winds (in the northern hemisphere) are seasonally constant and contribute to the aridity of South West Asia and the Middle East.
- Topography – When air is forced upward by a mountain range it will cool adiabatically (A to B) at the saturated adiabatic rate once the dew point is reached (B to C) with possible precipitation. On the leeward side of the mountain, the same air descends (C to D) warming at the dry adiabatic rate and hence the descending air is warmer at corresponding altitudes compared to the ascending air. Hence, a warmer, drier wind blows over the lands to the leeward side, provided that the ascent is sufficient to reach the dew point temperature.
- Cold ocean currents – Onshore winds blowing across a cold ocean current close to the shore will be rapidly cooled in the lower layers (up to 500 m). Mist and fog may result as found along the coasts of Oman, Peru and Namibia, but the warm air aloft creates an inversion preventing the ascent of air and hence there is little or no precipitation. As this air moves inland it is warmed and hence its humidity reduces.

The majority of semi-arid and arid regions are located between latitudes 25 and 35 degrees, where high pressures cause warm air to descend, resulting in dry, stable, air masses. Aridity caused by orographic aridity is common in North and South America, where high mountain ranges extend perpendicularly to the prevailing air mass movements. As described above, these air masses are cooled as they are forced up mountains, thus reducing their water-holding capacity. Most of the moisture is precipitated at the high elevations of the windward slopes. The relatively dry air masses warm as they descend on the leeward side of the mountain ranges, thus increasing their water-holding capacity and reducing the chance of any precipitation. This orographic aridity is called the rain shadow effect (Dick-Peddie, 1991).

The semi-arid and arid conditions of central Asia are caused by their position in the continent. As the distance from oceans increases, the chance of encountering moisture-laden air masses reduces. Cold ocean currents cause the coastal arid regions of Chile, Peru and the interior part of northern Argentina, where cold ocean currents in close proximity to the coast supply dry air that comes on shore, but as the mass is forced up the mountain sides there is no moisture to be lost as the air mass cools.

2.2.2 Climate areas

Logan (1968) distinguished the four areas as subtropical, continental interior, rain shadow and cool coastal arid lands, with the following definitions (Agnew and Anderson, 1992):

- Subtropical areas (e.g., Sahara, Arabia, Sonora, Australia and Kalahari) are characterized by anticyclonic weather, producing clear skies with high ground temperatures and a marked nocturnal cooling. The climate has hot summers and mild winters so the seasonal contrasts are evident with rare winter temperatures down to freezing. Convective rainfalls only develop when moist air invades the region.
- Continental interior areas (e.g., arid areas of Asia and western USA) have large seasonal temperature ranges from very cold winters to very hot summers. Snowfall can occur, however, its effectiveness may be reduced by ablation as it lies on the ground through winter. Rainfall in the summers is unreliable and can occur as violent downpours.
- Rain shadow areas (leeward sides of mountain ranges such as the Sierra Nevada, the Great Dividing Range in Australia and the Andes in South America) occur where conditions are diverse but are not as extreme as the continental interior areas.
- Cool coastal areas (e.g., Namib, Atacama and the Pacific coast of Mexico) have reasonably constant conditions with a cool humid environment. When temperature inversions are weakened by moist air aloft, thunderstorms can develop.

2.2.3 What are deserts?

Deserts implies aridity, however desert is a less precise term. There is no worldwide agreement as to what constitutes arid land and as to what gradations occur within the concept of arid (Dick-Peddie, 1991). Shmida (1985) equated extremely arid environments with extreme deserts, environments with deserts or true deserts and semi-arid environments with semi-deserts. Others equated arid and semi-arid environments with semi-deserts, which results in equating extremely arid environments with deserts (true deserts). Other authors use isohyets of annual precipitation to place limits on the various xeric zones. In most instances, the world's deserts tend to be located in Meigs' arid and extremely arid homoclimates.

2.3 HYDROLOGY

2.3.1 Rainfall

Precipitation includes rainfall, snowfall and other processes by which water falls to the land surface, such as hail and sleet (Chow et al., 1988). The formation of precipitation requires the lifting of an air mass in the atmosphere so that it cools and some of its moisture condenses. In arid environments, the processes leading to aridity tend to prevent cooling through maintaining air stability, creation of inversions or through the warming of the atmosphere resulting in the lowering of the humidity. The influence of these processes depends upon the atmospheric conditions. When rainfalls do occur they can be intense and localized downpours as moist air breaks through.

Slatyer and Mabbutt (1964) point out that the primary feature of precipitation in arid areas is the high variability of the small amount received, thus it is not uncommon for the standard deviation of the mean annual rainfall to exceed the mean value. They also point out that, in most arid regions, precipitation characteristics follow somewhat similar patterns, reflecting high variability in time and space of individual storms, seasonal rainfall, and annual and cyclical totals.

Schick (1988) discusses the immense temporal variability of rainfall and the very high intensities in hyper-arid areas of the world. In typical cloudbursts in the extreme desert areas, the transition between total dryness and full-blast rain is near instantaneous, with the first few minutes of the rainfall having intensities in excess of 1 mm/min. The excessive intensities in hyper-arid areas seem to be associated with relatively high temperatures, and are therefore the result of convective processes. The convective storms tend to form at preferred distances from each other, as opposed to being randomly scattered in space. Sharon (1981) found that the convective storms in the extremely arid Namib Desert had preferred distances of around 40–50 km and 80–100 km. The rain fronts are often sharply defined both in the direction of cell movement as well as laterally. Sharon (1972) reported cases where the velocity of rain cells in the extreme desert was found to vary from near zero (a stationary cloudburst) to several tens of kilometres per hour. The lateral boundaries of moving cells tend to be sharp. There are also widespread rainfalls that cover vast desert areas with lower intensity but a relatively high quantity of rain.

Goodrich et al. (1990) studied the impacts of rainfall sampling on runoff computations in arid and semi-arid areas of the south-western US. This study concluded that the appropriate rainfall-sampling interval for arid land watersheds depends on many factors, including the temporal pattern of the rainfall intensity, watershed response time and infiltration characteristics. The study recommended that either breakpoint rainfall data of data sampled at uniform time increments be used for watersheds with equilibrium times smaller than about 15 minutes and that a maximum interval of 5 minutes be used for basins that respond more slowly. Using a physically based, rainfall–runoff model (KINEROS), Woolhiser et al. (1990), found that the outflow hydrographs were more sensitive to the rainfall input than to the model parameters. Fenreira (1990) used Opus, an agricultural ecosystem model with an infiltration-based hydrology option, to simulate field responses in arid and semi-arid areas to rainfall inputs of various time intervals. The results using synthetic rainfall data from the statistical analysis of rainfall data from watersheds in the south-western US showed a strong sensitivity of runoff predictions to the time interval of input rainfall data.

2.3.2 Infiltration

In arid lands the physical process of soil formation is active, resulting in heterogeneous soil types, having properties that do not differ greatly from the parent material, and having soil profiles that retain their heterogeneous characteristics (Elgabaly, 1980). Soils in arid lands may contain hardened or cemented horizons known as pans and classified according to the cementing agent, such as gypserious, calcareous, iron and so on. The extent to which the horizons affect infiltration and salinization depends upon their thickness and depth of formation as they constitute an obstacle to water and root penetration. In salt mediums, salt crusts can form at the surface under specific conditions.

Large quantities of soluble salts cause the coagulation of clay particles. Enrichment of the soil in sodium salts modifies the soil structure, because of the dispersion and swelling properties of sodic clays generally formed, and the soils become impermeable. Changes in soil structure due to the action of different salts, has an important influence on the behaviour of soil under irrigation and drainage. Salt-affected, arid, sodic soils have a very loose surface structure, making them susceptible to wind and water erosion.

The determination of the effect of an impervious area is particularly important. An impervious area increases the volume of runoff and increases the velocity of the water, both of which tend to increase peak flows. However, the effect of an increased impervious area depends upon its location in the basin and the 'connectedness' of the impervious area to the channel. Runoff from the impervious areas not directly connected to the channel system must flow over pervious areas, and thus contribute less runoff.

2.3.3 Runoff and flooding

In arid and semi-arid regions, flash floods are caused by high intensity, short duration storms with a high degree of spatial variability. Runoff hydrographs from these storms typically exhibit very short rise times, even for large catchments (Goodrich et al., 1990).

2.3.3.1 Overland flow

When the rainfall rate exceeds the infiltration capacity, and sufficient water ponds on the surface to overcome surface tension effects and fill small depressions, Hortonian overland flow begins (Woolhiser et al., 1990). When viewed from a micro-scale, overland flow is a three-dimensional process, however, at a larger scale it can be viewed as a one-dimensional flow process in which the flux is proportional to some power of the storage per unit area, as $Q = ah^m$, where Q is the discharge per unit width, h is the storage of water per unit area (or depth if the surface is a plane), a and m are parameters related to slope, surface roughness and whether the flow is laminar or turbulent. The continuity equation for flow is expressed through the kinematic-wave assumption.

The kinematic assumption requires only that discharge be some unique function of the amount of water stored per unit of area; it does not require sheet flow (Woolhiser et al., 1990). The kinematic-wave formulation is an excellent approximation for most overland flow conditions (Woolhiser and Liggett, 1967 and Morris and Woolhiser, 1980).

Lane et al. (1978) studied the partial area response (variable source area response) on small semi-arid watersheds. This refers to the response of a watershed when only a portion of the total drainage area is contributing runoff at the watershed outlet or point of interest. The generation of overland flow on portions of small semi-arid watersheds was analyzed using three methods: an average loss rate procedure, a lumped-linear model and a distributed-nonlinear model (kinematic wave). The results showed that significant errors in estimating surface runoff and erosion rates are possible if a watershed is assumed to contribute runoff uniformly over the entire area, when only a portion of the watershed may be contributing.

2.3.3.2 Floods and channel routing

From a geomorphology viewpoint, the response of fluvial systems to flood discharges depends in large part on the amount of time that has passed since the major climatic perturbation switched the mode of operation of hill slopes (mid-slopes and foot slopes) from net aggradation to net degradation (Bull, 1988). During the early stages of hill slope stripping, the amount of available sediment is so large that intense rainfall–runoff events cause debris flows and accelerate valley floor aggradation. During later stages, an opposite result occurs, when major rainfall events accelerate the removal of the remaining sediment on the hill slopes, thereby causing still larger increases of stream power relative to resisting power. The resulting degradation cuts through the valley fill to bedrock. Unsteady, free, surface flow in channels can also be represented by the kinematic-wave approximation to unsteady, gradually varied flow (Chow et al., 1988).

2.3.3.3 Transmission losses

Many semi-arid and arid watersheds have alluvial channels that abstract large quantities of stream flow, called transmission losses (Lane, 1982, 1990; Renard, 1970). These losses are important in the determination of runoff because water is lost as the flood wave travels downstream. The transmission losses are an important part of the water balance because they support riparian vegetation and recharge local aquifers and regional groundwater (Renard, 1970). Procedures to estimate transmission losses range from inflow-rate-loss equations, to simple regression equations, to storage-routing as a cascade of leaky reservoirs and to kinematic-wave models incorporating infiltration (Woolhiser et al., 1990). Stream channels also transport water across alluvial fans from mountain fronts to lower portions of the watersheds. These channels are unstable and are variable in time and space; however they retain their ephemeral character and thus transmission losses can exhibit their influence on flood peaks, water yield and groundwater recharge just as for ephemeral stream channel networks (Lane, 1990).

From the viewpoint of flood routing and transmission losses, the main difference between ephemeral stream channel networks forming the drainage patterns in watersheds and ephemeral channel segments transverse alluvial fans is due to the nature of their structure and linkage (Lane, 1990). In watersheds, the channel systems tend to be dendritic, with the main channels collecting tributary inflow in the downstream direction. On alluvial fans, the channel segments tend to be singular or bifurcating in the downstream direction. On alluvial fans, there are usually no tributary inflows; however, channels can split or diverge resulting in tributary outflows in the downstream direction. As pointed out by Lane (1990), in spite of the differences, many of the same flow processes occur in watersheds and on alluvial fans; therefore, methods that have been developed to consider stream flow and transmission losses in individual stream channel segments can be applied to both ephemeral streams in watersheds and to ephemeral stream segments transversing alluvial fans.

2.3.4 Erosion and sediment transport

Water is the most widespread cause of erosion. It can be classified into sheet erosion and channel erosion. Sheet erosion is the detachment of land surface material by raindrop impact and the thawing of frozen grounds and its subsequent removal by overland flow (Shen and Julian, 1993). The transport capacities of thin overland flow or sheet flow, increases with field slope and flow discharge per unit width. As the sheet flow concentrates and the unit discharge increases, the increased sediment transport capacity scours micro-channels, also referred to as rills. Rill erosion is the removal of soil by concentrated sheet flow. Surface erosion begins when raindrops hit the ground and detach soil particles by splash (Shen and Julien, 1993).

The rate of upland erosion of the soil bed, e, can be composed of two components: 1) soil erosion by the splash of rainfall on bare soil and 2) hydraulic erosion (or deposition). Hydraulic erosion is due to the interplay between the shear force of water on the loose soil bed and the tendency of soil particles to settle under the force of gravity.

Channel erosion includes both bed and bank erosion, which can be very significant in alluvial channels. Sediment transport capacity is generally proportional to the water discharge and channel slope. The sediment transport capacity varies inversely with the bed sediment size. Sediment transport simulation for channels can be nearly the same as for upland areas. The difference is that splash erosion is neglected for channel erosion (Woolhiser et al., 1990).

Wind erosion can be important in semi-arid and arid areas. The rate of wind erosion depends upon the particle size distribution, wind velocity, soil moisture, surface roughness and vegetative cover. Chepil and Woodruff (1954) proposed an empirical equation for estimating the rate of wind erosion. This equation provides a rough estimate of wind erosion rates.

2.4 URBAN WATER MANAGEMENT

Three major challenges to urban water management, particularly in arid and semi-arid regions in the world are the demographic challenge, the economic challenge and the urbanization challenge. A study initiated by UNESCO – Problem assessment and strategic urban planning on urban water management – focused on the Middle East and Central Asia region. This study was performed at the Regional Centre on Urban Water Management (RCUWM) based in Tehran, Iran. The countries in this region are facing a large demographic challenge. Population growth in Islamic countries has been greater compared to the rest of the world. In Muslim societies, annual growth rates have been over 2.0% and sometimes over 3.0% (Figueres, 2005) compared to annual growth rates of 1.85% for the total population on earth between 1965 and 1990. Disparities between countries in this region include not only the total population, population density and percentage of urban population, but also in the income per capita. As an example, there are the very rich Gulf States and also some of the poorest countries on earth. Oil and tourism are the two industries that dominate the economic life of most countries of this region. Open unemployment in the Arab region has been estimated to be around 15%. Because of the demographic and economic challenge, the urbanization challenge is growing. As in many other parts of the world, the provision of housing, infrastructure and urban services cannot keep up with the rapid urbanization.

The UNESCO study of the Middle East and Central Asia identified several main issues:

- water resources scarcity
- aged infrastructure and unaccounted for water
- lack of sewerage and wastewater treatment
- poor solid waste collection and disposal

- poor management and institutional framework
- low cost recovery
- high subsidies
- negative effects of wars and natural disasters
- presence of water-related illnesses
- lack of integrated planning of municipal infrastructure
- urgent need for capacity building of staff.

This study showed that the urban water problems in the Middle East and Central Asia, fundamentally, are no different than other regions of the world (Figueres, 2005).

Throughout the world the major emphasis for urban water has been placed on the development of infrastructure. However, infrastructure alone clearly is not the answer to supply sustainable urban water needs. Based upon the UNESCO study the following interventions should be addressed when planning urban water infrastructure development (Figueres, 2005):

- develop an integrated urban planning strategy
- develop a water demand management strategy and campaign
- improve wastewater collection and treatment systems
- encourage reuse and recycling of treated wastewater
- renovate or replace aging infrastructure by adapted systems
- install detection systems in water supply networks and reduce unaccounted for water
- promote water conservation (e.g., reduce subsidies, promote block tariffs, install meters, public awareness)
- improve solid waste management and encourage recycling
- create an enabling institutional framework
- create a legal and regulatory framework and implement laws
- carry out risk assessments for unpredictable events
- reinforce cooperation between public and private actors at all levels in planning, design, implementation, operation, maintenance and management of urban infrastructure and services where appropriate
- give high priority to training and capacity building for sustainability.

Chapter 3

Integrated water supply management in arid and semi-arid regions

The complete conventional water supply system includes all the system components to develop drinking water and distribute it to the users, and all the system components to collect and treat the wastewaters. In arid and semi-arid regions, the conventional water supply system is similar to those found in other climates. This chapter mainly focuses on technologies that are available and applicable to arid and semi-arid climates for developing water supplies which are typically not considered as part of conventional systems. These technologies include water reuse and artificial recharge systems, desalination of seawater, water transfers and rainfall harvesting.

3.1 OVERALL SUBSYSTEM COMPONENTS AND INTERACTIONS: CONVENTIONAL SYSTEMS

This section very briefly addresses both the conventional drinking-water system and the wastewater and collection system as part of the overall conventional water supply system. These conventional systems are discussed in detail in many textbooks and other publications, so they are not addressed herein.

3.1.1 Drinking-water supply systems

Conventional water supply systems include the following components: water sources (groundwater and surface sources); raw water pumping and transmission; water treatment; high service pumping; and water distribution (as illustrated in Figure 3.1). A detailed discussion of the various components is beyond the purposes of this book. Water distribution systems are discussed in many publications including Mays (2000b, 2002 and 2004b).

Loss of water in supply networks (water distribution systems) is a worldwide problem, but it is even more pronounced in many countries, particularly developing countries in arid and semi-arid regions. Examples (see Post, 2006) include Jordon (50% losses), Palestine (from 36 to 60% losses) and Lebanon (around 50% losses). Case study I, on Mexico City, reports 40% losses in the distribution system and the poor quality of the water due to contamination caused by erratic low pressures in the system. Case study V, on Cairo, Egypt, also discusses similar problems.

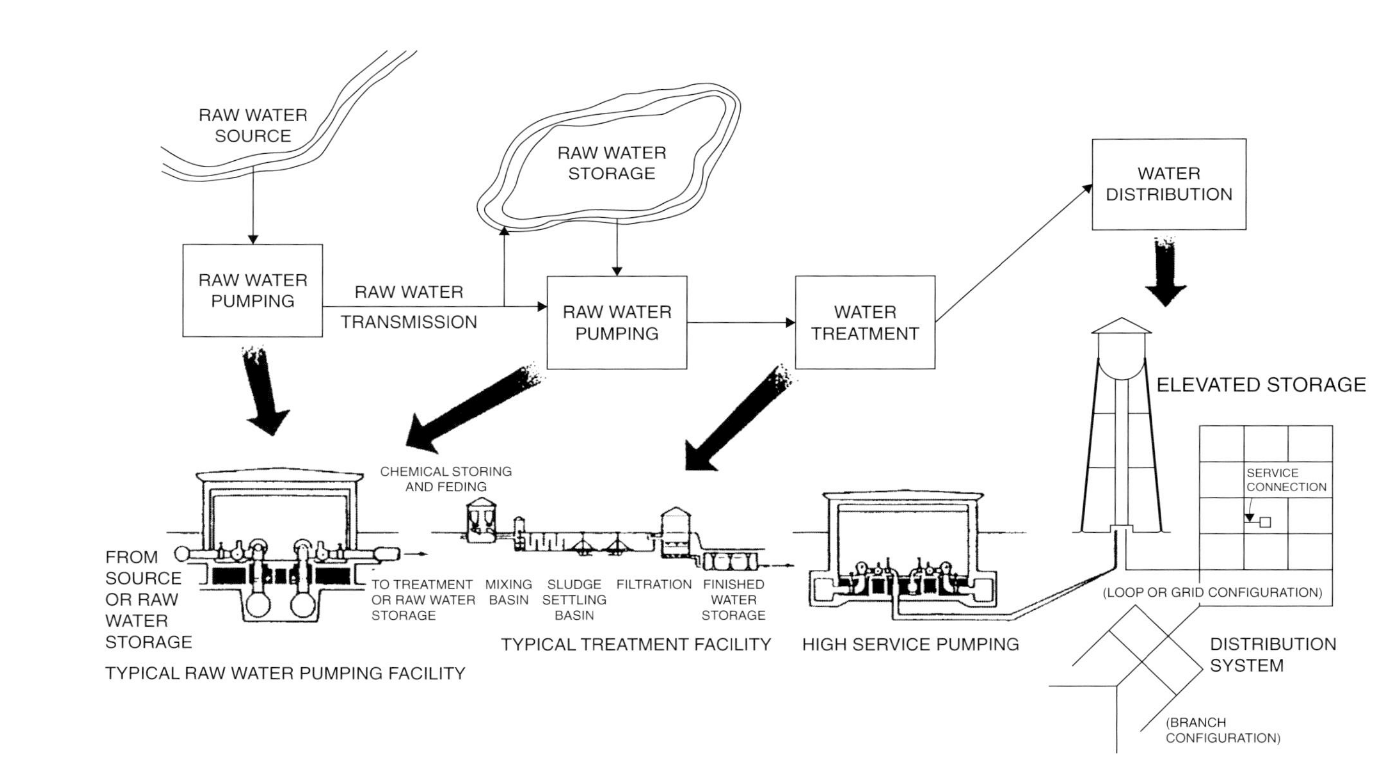

Figure 3.1 Conventional layout of water sources, pumping, transmission, water treatment, water distribution

Source: Cullinane, 1989

Another major problem is the erratic pressure caused by inadequate supplies and operational problems that allow flows of contaminated groundwater and/or sewage into the water distributions network. These flows enter the distribution system when the pressure is low through cracks, faulty valves and other appurtenances. As a result, many of the cities have systems that supply water but it cannot be consumed without boiling. The Mexico City case study discusses this in further detail.

3.1.1.1 Reliability measures

To ensure delivery of finished water to users, water distribution systems must be designed and operated to accommodate a range of expected emergency loading conditions. These emergency conditions may generally be classified into three groups: broken pipes, fire demands, and pump and power outages. Each of these conditions must be examined with an emphasis on describing its impact on the system, developing relevant measures of system performance and designing into the system the capacity required to handle the emergency condition with an acceptable measure of reliability. Reliability is usually defined as the probability that a system performs its mission within specified limits for a given period in a specified environment.

The reliable delivery of water to the user requires that the water distribution system be designed to handle a range of expected emergency loading conditions. These emergency conditions can be classified as:

- fire demands
- broken links
- pump failure
- power outages
- control valve failure
- insufficient storage capability.

A general application methodology that considers the minimum cost and reliability aspects must consider each of these emergency conditions. These conditions should be examined within the methodologies to:

- describe their importance to the system
- develop relevant measures of system performance
- design into the system the capacity for the emergency loading conditions with an acceptable measure of reliability.

The 'failure' of water distribution networks can be defined as either the pressure and/or flow falling below specified values at one or more nodes within the network. Under such as a definition, there are two major modes of failure of water distribution networks:

- performance failure, i.e. demand on the system being greater than what the system can produce and/or pressures not being satisfied because of limited supplies at various times resulting in erratic pressure
- component (mechanical) failure, which includes the failure of individual network components, e.g., pipes, pumps and valves.

Both these modes of failure have probabilistic bases which must be incorporated into any reliability analysis of networks.

3.1.1.2 Performance indicators

Because of the requirement for the availability and good quality of drinking water, the concept of performance indicators has received considerable attention in Europe and many other countries (Alegre, 2002). Performance indicators are measures of the efficiency and effectiveness of the water utilities with regard to specific aspects of the utility's activity and of the system's behaviour. Efficiency is a measure of the extent to which the resources of a water utility are utilized optimally to produce the service, while effectiveness is a measure of the extent to which the targeted objectives (specifically and realistically defined) are achieved. Each performance indicator expresses the level of actual performance achieved in a certain area and during a given period, allowing for a clear-cut comparison with targeted objectives and simplifying an otherwise complex analysis. According to the definition recommended by the International Water Association (IWA), (Alegre et al., 2000) any performance indicator shall be defined as a ratio between variables of the same nature (e.g., %) or of different natures (e.g., $/m^3; litres/service connection).

3.1.2 Wastewater and collections systems

Wastewater collection (sewerage) systems have the function of conveying wastewater produced from residents, institutions, and commercial and industrial establishments, and, in some locations, stormwater to points of wastewater treatment and disposal. Wastewater collection systems typically consist of a network of pipes and pumping stations for transporting wastewater to a final destination.

Sewerage systems in many developing countries in arid and semi-arid regions are limited. In Palestine about 24% of the total population is served by a central public urban sewerage system with around 73% of households using cesspits and septic tanks, and 3% having no sanitation systems (Post, 2006). In Lebanon about 60% of households are connected to sewerage systems and many towns and villages do not have a wastewater infrastructure except for household pits (Post, 2006). In Turkey about 50% of the urban population is connected to sewerage systems (Post, 2006).

Conventional wastewater treatment facilities are based on an analysis of:

- physical, chemical and biological characteristics of the wastewater
- the quality that must be maintained in the environment to which the wastewater is to be discharged for the reuse of wastewater
- the applicable environmental standards or discharge requirements that must be met (Tchobanoglous, 1996).

Technologies for urban wastewater treatment as applied to the specifics found in warm climates are presented in Von Sperling and Chernicharo (2005). Because most developing countries are in warm climates, they provide a special overview of the real-life situation where there is a demand for simple, economical and sustainable solutions (Von Sperling, 1996). Temperature has a decisive role in some treatment processes, such as those based upon natural processes and non-mechanized systems. Climate has a large influence on biological treatment processes. Nitrification is influenced by temperature. In warm climates it is likely to take place even under unfavourable conditions and at low sludge ages in activated sludge reactors. Warmer temperatures require less land, increased removal efficiencies and enhanced conversion processes. In warmer climates, the use of certain treatment processes is more feasible, such as the use of anaerobic reactors for diluted wastewaters from domestic sewage. Areas required for stabilization ponds in warm climate regions are three to four times smaller than in temperate regions.

In many countries in arid and semi-arid regions the availability of wastewater treatment facilities is in short supply. As an example, in Palestine today less than 6% of the total population is served by wastewater treatment, and the only large-scale plant that works went into operation in 2000 (Post, 2006). This plant is an extended aeration treatment process that currently serves around 50,000 people, and will eventually serve 100,000 people. In most of Lebanon and Palestine, most of the wastewater collected by sewerage systems is discharged, with little or no treatment, into nearby rivers, wadis, the sea, or on open land or underground. This practice obviously pollutes the environment and poses serious human health risks.

3.2 WATER RECLAMATION AND REUSE

Water reclamation and reuse has become an attractive option for conserving and extending available water supplies in arid and semi-arid regions, as many urban areas in these regions approach and reach the limits of their available water supplies. Water reuse can also provide urban areas with an opportunity for pollution abatement when effluent discharges are replaced to sensitive surface water. Water reclamation and nonpotable reuse only require conventional water and wastewater treatment technologies that are widely practised and available in countries throughout arid and semi-arid regions of the world.

The technical issues involved in planning a water reuse system include (US EPA and USAID, 1992):

- The identification and characterization of potential demands for reclaimed water.
- The identification and characterization of existing sources of reclaimed water to determine their potential for reuse.
- The treatment requirements for producing a safe and reliable reclaimed water that is suitable for its intended applications.
- The storage facilities required to balance seasonal fluctuations in supply with fluctuations in demand.
- The supplemental facilities required to operate a water reuse system, such as conveyance and distribution networks, operational storage facilities and alternative disposal facilities.
- The potential environmental impacts of implementing water reclamation.

Figure 3.3 illustrates three alternative configurations for water reuse systems and Figure 3.4 shows the typical water reclamation plant process for urban reuse.

The Mediterranean region is one of the most affected water-short regions of the world; improved water-demand management and development of new water resources is desperately needed. In this region, wastewater is in most cases, inadequately

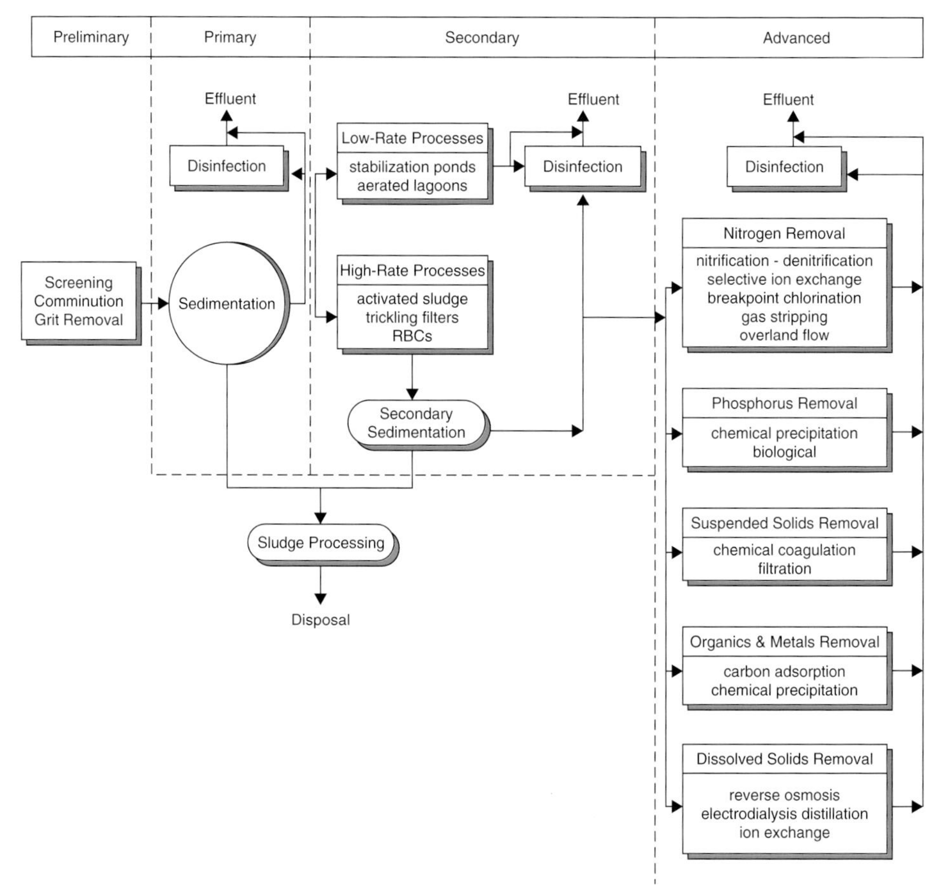

Figure 3.2 Generalized flow sheet for municipal wastewater treatment

Source: US EPA, 2004 as adapted from Pettygrove and Asano, 1985

managed and treated, which in turn leads to the deterioration of existing freshwater sources and even the Mediterranean Sea (Post, 2006). An EU-funded project, 'Efficient Management of Wastewater, Treatment and Reuse in the Mediterranean Countries', encourages reuse-oriented wastewater management in four target countries – Jordan, Palestine, Lebanon and Turkey. Post (2006) provides an overall comparison of the different specific problems within the targeted countries. The conclusions are that because of the existing and predicted water deficits in the region, there needs to be an increase in water use efficiency and alternative water sources, such as reclaimed wastewater, must be considered as an option. A large effort will be required to develop a cultural acceptance, as it is a key problem. There needs to be a successful demonstration of wastewater treatment and reuse technologies to convince the people of the benefits.

3.3 MANAGED AQUIFER RECHARGE (MAR)

Managed aquifer recharge (MAR) has the potential to be a contributor to the UN Millennium Goals for water supply, especially in arid and semi-arid areas. MAR can be regarded as one method to manage urban water resources in conjunction with others.

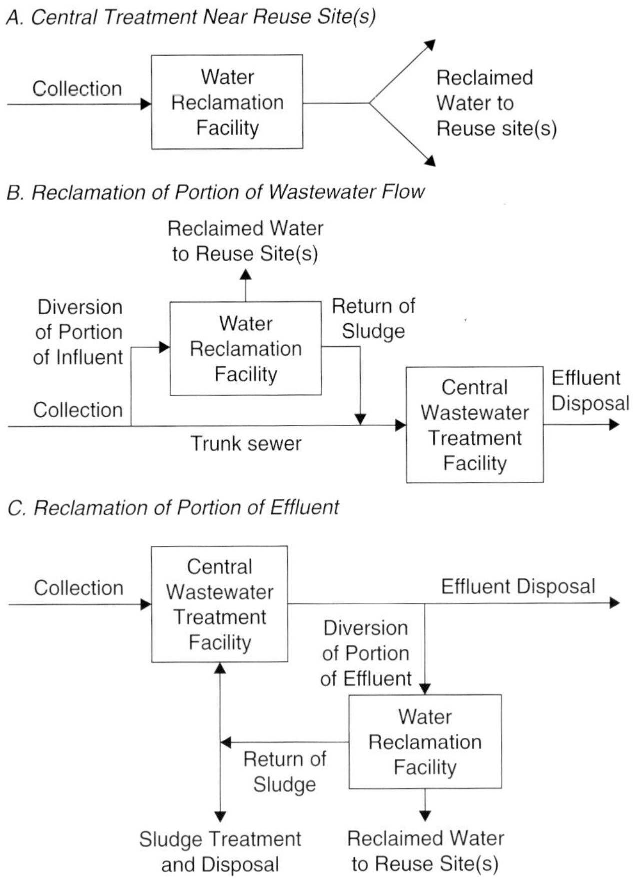

A. Central Treatment Near Reuse Site(s)

Collection
Water Reclamation Facility
Reclaimed Water to Reuse site(s)

B. Reclamation of Portion of Wastewater Flow

Reclaimed Water to Reuse Site(s)
Diversion of Portion of Influent
Water Reclamation Facility
Return of Sludge
Collection
Trunk sewer
Central Wastewater Treatment Facility
Effluent Disposal

C. Reclamation of Portion of Effluent

Collection
Central Wastewater Treatment Facility
Effluent Disposal
Diversion of Portion of Effluent
Water Reclamation Facility
Return of Sludge
Sludge Treatment and Disposal
Reclaimed Water to Reuse Site(s)

Figure 3.3 Configuration alternatives for water reuse systems. (A) Wastewater treatment facility located near major users of the reclaimed water for purposes of economy. (B) When trunk sewer passes through an area of significant potential reuse a portion of the flow can be diverted to a new reclamation facility to serve that area. Sludge from that facility is returned to sewer which could be deleterious to sewer trunk and downstream treatment facility. (C) Effluent outfall passing through a potential reuse area could be tapped for some or all effluent and provide additional treatment to meet reclaimed water quality standards

Source: US EPA and USAID, 1992, 2004

3.3.1 Artificial recharge

Artificial recharge is becoming an integral part of integrated urban water management for sustainable water supplies, particularly in arid and semi-arid regions, and particularly in the south-western United States. Aquifer recharge has many advantages over conventional surface water storage, especially in arid and semi-arid regions. These include negligible evaporation losses, limited vulnerability to secondary contamination by animals and/or humans and no algae blooms resulting in decreasing surface water quality (Crook, 1998). The flow of water through the subsurface provides soil-aquifer treatment (SAT), along with aquifers providing seasonal and/or longer storage.

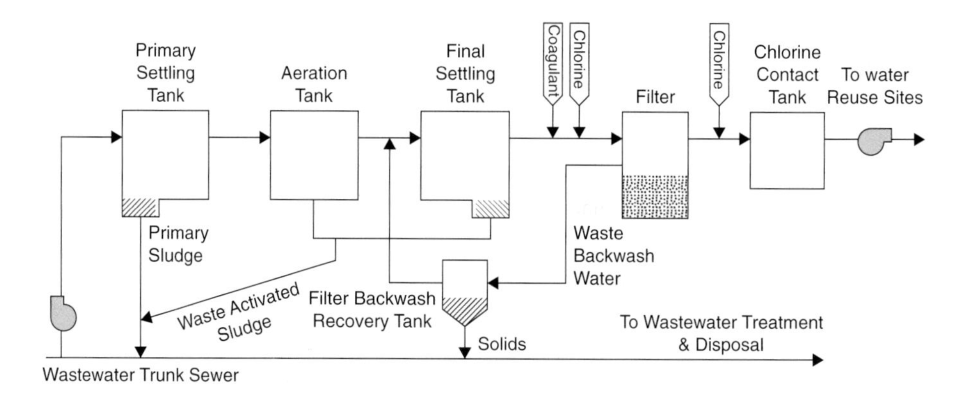

Figure 3.4 Typical water reclamation plant processes for urban reuse

Source: US EPA and USAID, 1992, 2004

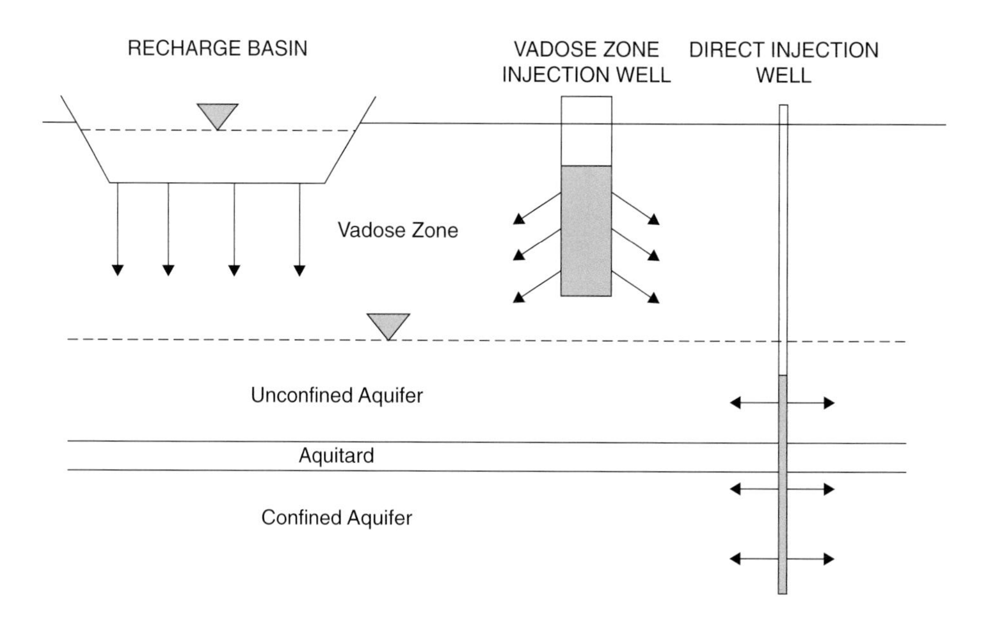

Figure 3.5 Methods for aquifer recharge

Source: Fox, 1999 and US EPA, 2004

Groundwater recharge of reclaimed municipal wastewater is an approach to water reuse with the following purposes (Asano, 1985, 1999; Bouwer, 1978; Todd and Mays, 2005):

- to reduce, stop or even reverse declines of groundwater levels
- to protect underground freshwater in coastal aquifers against saltwater intrusion
- to store surface water, including flood or other surplus water, and reclaimed municipal wastewater for future use.

Table 3.1 Major characteristics of aquifer recharge of reclaimed municipal wastewater (Fox, 1999)

	Recharge basins	*Vadose zone injection wells*	*Direct injection wells*
Aquifer type	Unconfined	Unconfined	Unconfined or confined
Pre-treatment requirements	Low technology	Removal of solids	High technology
Estimated major Capital costs US$	Land and distribution system	\$25,000–75,000 per well	\$500,000–1,500,000 per well
Capacity	1000–20,000 m^3/ha-d	1000–3000 m^3/well-d	2000–6000 m^3/well-d
Maintenance requirements	Drying and scraping	Drying and disinfection	Disinfection and flow reversal
Estimated life cycle	>100 Years	5–20 Years	25–50 Years
Soil-aquifer treatment	Vadose zone and saturated zone	Vadose zone and saturated zone	Saturated zone

Figure 3.5 illustrates the three technologies – percolation or recharge basins, vadose zone injection wells and direct injection wells – for aquifer recharge of reclaimed municipal wastewater. The major characteristics of these technologies are summarized in Table 3.1. Each of the three technologies provides soil-aquifer treatment (SAT) as noted in the table. The aquifer recharge basins and vadose zone injection wells provide SAT in both the vadose and saturated zones; whereas the direct injection wells provide SAT in the saturated zone and in the unconfined and/or confined zones. The percolation or recharge basin is the most common and widely accepted method (Fox, 1999).

3.3.2 Recharge basins

A recharge basin consists of five major components (see Figure 3.6): 1) the pipe line that carries the treated effluent from the wastewater treatment plant; 2) percolation (infiltration) basins, where the treated effluent infiltrates into the ground; 3) the soil immediately below the infiltration basins (vadose zone); 4) the aquifer, where water is stored for a long duration; and 5) the recovery well, where water is pumped from the aquifer for potable or non-potable reuse. A glossary of terms is provided at the beginning of this report.

Figure 3.7 illustrates the SAT system dynamics, which can be described by the inputs to the system, the state of the system and the outputs from the system. Inputs to the system include: 1) the soil type; 2) the water quality; 3) the operation of the SAT system; and 4) the environment, where each input affects the state of the system. The state of an SAT system includes: 1) the soil moisture profile within the vadose zone; 2) the level of oxygen; 3) algal growth; and 4) soil hydraulic conductivity. The state of the system controls the residence time of water in the vadose zone and the level of microbial activity. Of particular importance is the level of oxygen which is related to the microorganism distribution and oxygen-demanding substrates. Algal growth is related to the clogging layer formation on the soil surface and effective soil hydraulic conductivity. The soil moisture is directly affected by inputs – soil type and the environment – and indirectly by the water quality, which affects algal growth and the soil hydraulic conductivity. The oxygen in the vadose zone is affected by the soil moisture profile, the

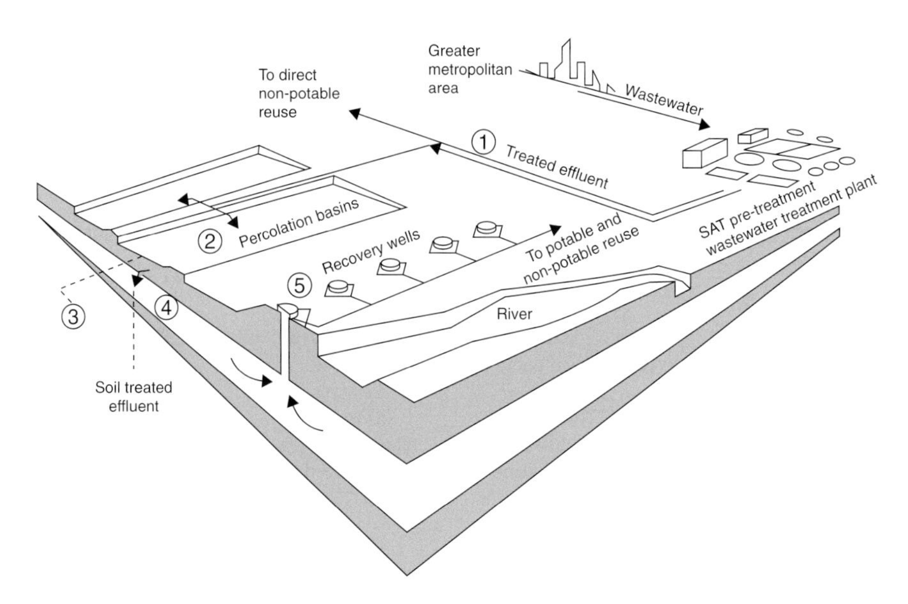

Figure 3.6 Recharge basin and the components

Source: Arizona State University et al., 1998

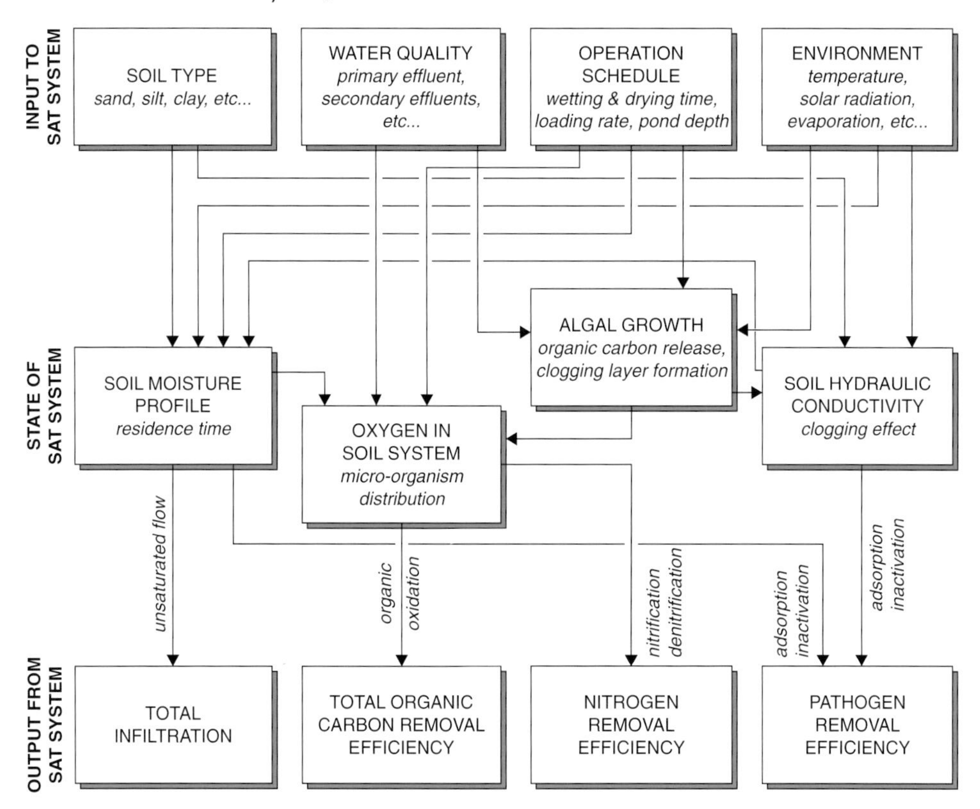

Figure 3.7 Soil-aquifer treatment (SAT) system dynamics

Source: Arizona State University et al., 1998

water quality of treated effluent, the operation schedule and the algal growth. Algal growth is affected by the water quality of the treated effluent, the operation schedule and the environment. Soil hydraulic conductivity is affected by the algal growth, the soil type and the environment.

The critical outputs of the system are the total infiltration, the total organic carbon removal efficiency, the nitrogen removal efficiency and the pathogen removal efficiency. The total organic carbon removal efficiency is primarily affected by the amount of oxygen in the system through the organic oxidation process. The nitrogen removal efficiency is depends upon the amount of oxygen and the availability of organic carbon through the nitrification–denitrification process. The pathogen removal efficiency depends upon the soil moisture profile and the soil hydraulic conductivity through the adsorption–inactivation process.

The major purification processes occurring in the soil-aquifer system are: filtration, chemical precipitation/dissolution, organic biodegradation, nitrification, denitrification, disinfection, ion exchange and adsorption/desorption. As part of a water reclamation process, soil-aquifer treatment (SAT) is a key step to polishing the water. This step provides for:

- mechanical filtration of suspended particles
- biological processes (e.g., breakdown of organics, nitrification–denitrification)
- physical-chemical retention of inorganic and organic dissolved constituents (e.g., phosphorous, potassium, trace elements), from the biologically treated wastewater.

Water reuse can be one of the most important strategies for meeting water quantity objectives in arid and semi-arid regions. Local groundwater supplies can be augmented by treated wastewater that re-enters the ground via managed infiltration practices or injection wells. The infiltration of wastewater effluents to augment local aquifers is accomplished in shallow infiltration basins, usually constructed in regions with permeable soils. The treated water percolates through the soil mantle and unconsolidated sediments, and then the vadose zone, to reach an unconfined aquifer. Although these activities are generally not located near production wells, waters that are added to local aquifers in this manner mix with the native groundwater and are eventually recovered and used again.

Figures 3.8 through 3.10 show three artificial recharge systems in Arizona.

3.4 DESALINATION

Desalination refers to any one of several processes that remove the excess salt and other minerals from water to obtain freshwater suitable for consumption or irrigation. Desalination is an unconventional method for water supply that is gaining wider use, particularly in arid and semi-arid regions of the world. Because of water scarcity in the Middle East, desalination of ocean water is becoming more common. It is also growing in the USA, North Africa, Spain, Australia and China. Table 3.2 is a list of the number of desalination units in the MENA (Middle East and North Africa) Region as reported by the Gulf Cooperation Council (GCC).

Figure 3.8 Pima Road recharge basin near Tucson, Arizona (See also colour plate 3)

Source: Courtesy of Central Arizona Project

Several Middle Eastern countries have energy reserves so great that they use desalinated water for agriculture. Saudi Arabia's desalination plants account for about 24% of total world capacity. Desalination meets 70% of Saudi Arabia's present drinking-water requirement, supplying major urban and industrial centres through a network of more than 2,300 miles of pipe network. The world's largest desalination plant is the Shoaiba Desalination Plant in Saudi Arabia. It uses multi-stage flash distillation, and it is capable of producing 150 million cubic metres of water per year. Several new desalination plants are planned, or under construction, which will bring the final total to almost 30 such facilities.

A new city, the Dubai World Central (DWC), is to be located in 'New Dubai' in the arid United Arab Emirates. DWC will host the Dubai World Central International Airport, a Logistic City, an Aviation City, an Exhibition Centre, an Enterprise Park and others, with a proposed residential population of over 988,000 and employment of over 608,000. Desalinated water will supply DWC with final potable and cooling water demands, respectively of 400,000 and 175,000 m^3/day.

The methods for desalination include the following:

3.4.1 Distillation

- Multi-stage flash (see Figure 3.12)
- Multiple-effect (see Figure 3.12)
- Vapour compression (see Figure 3.12)
- Evaporation/condensation.

3.4.2 Membrane processes

- Electrodialysis reversal
- Reverse osmosis (see Figure 3.11)

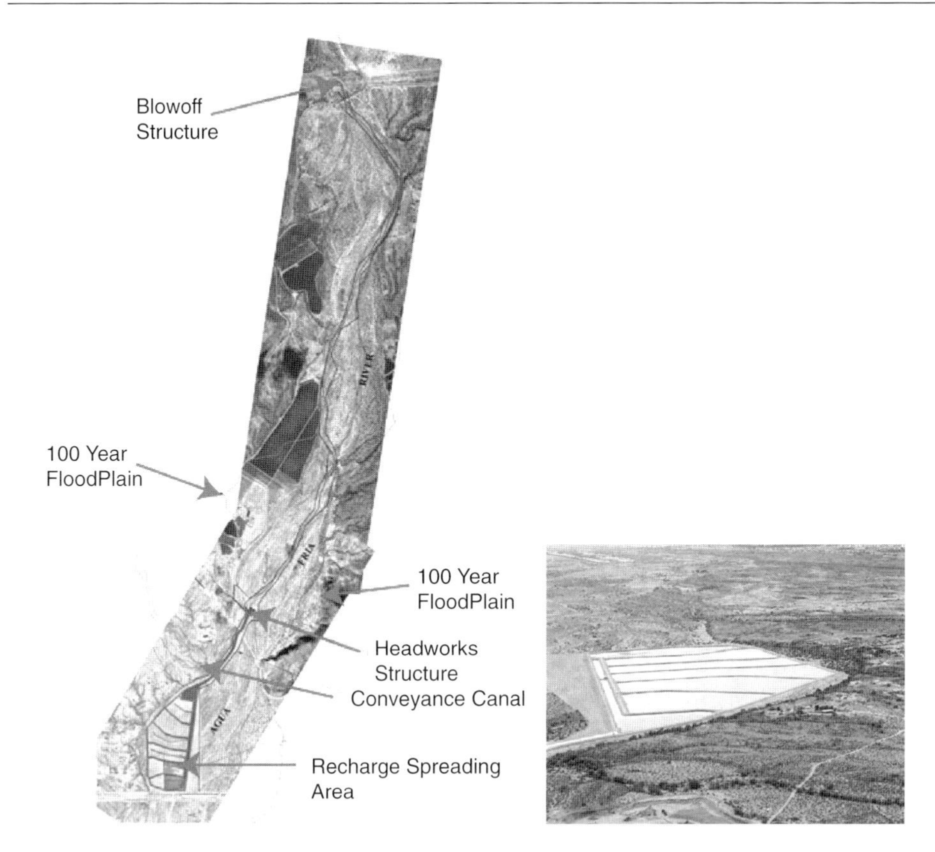

Figure 3.9 Agua Fria Recharge Project (AFRP) located in Agua Fria River, Peoria, Arizona. This project is located approximately four miles downstream of the New Waddell Dam (Lake Pleasant). The project was developed by the Central Arizona Water Conservation District (CAWCD) and in 2003 the City of Peoria purchased AFRP storage capacity for recharge to meet the demands of future growth as part of their water resources management goals. Two operational components include the four-mile river section used for recharge and conveyance of surface water downstream, and a constructed head structure to capture surface flow in the river and a canal to convey water downstream to the spreading basins (100 acres in area) (See also colour plate 4)

Source: Courtesy of the Central Arizona Project

- Nanofiltration
- Forward osmosis
- Membrane distillation.

Other methods include: freezing, geothermal, solar humidification, methane hydrate crystallization and high-grade water cycling.

The two leading methods are reverse osmosis (47.2% of installed capacity world-wide) and multi-stage flash (36.5%). (Source: 2004 IDA Worldwide Desalting Plants Inventory Report No 18; published by Wangnick Consulting). Electrodialysis is usually preferred for treating brackish groundwater.

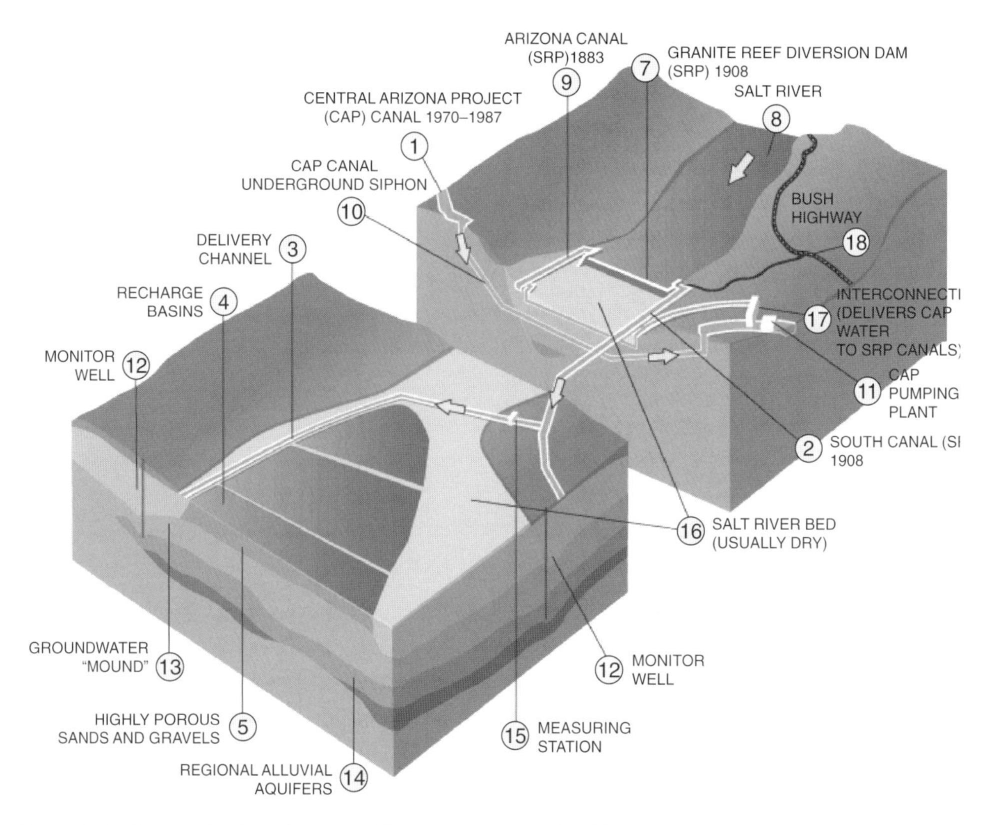

Figure 3.10 Granite Reef Underground Storage Project (GRUSP) operated by the Salt River Project (SRP), Arizona. This facility diverts water from the Granite Reef Diversion Dam near Phoenix, Arizona on the Salt River to seven recharge basins totalling 217 acres for the purpose of water banking. Actual recharge is approximately 100,000 ac-ft/yr with 200,000 ac-ft/yr permitted by the State of Arizona. This was the first major recharge facility in the state of Arizona and one of the largest in the US. (See also colour plate 5)

Source: Courtesy of the Salt River Project

Table 3.2 Desalination units in MENA (Middle East and North Africa) Region

Location	*Number of Units*	*Total Capacity (m^3/d)*
UAE	382	5,465,784
Bahrain	156	1,151,204
Saudi Arabia	2074	11,656,043
Oman	102	845,507
Qatar	94	1,223,000
Kuwait	178	3,129,588
GCC states total		**23,471,126**
Libya	431	1,620,652
Iraq	207	418,102
Egypt	230	236,865
Algeria	174	301,363
Tunisia	64	148,822
Yemen	66	132,897
Israel	n/a	149,594

Source: Arab Gulf Cooperation Council

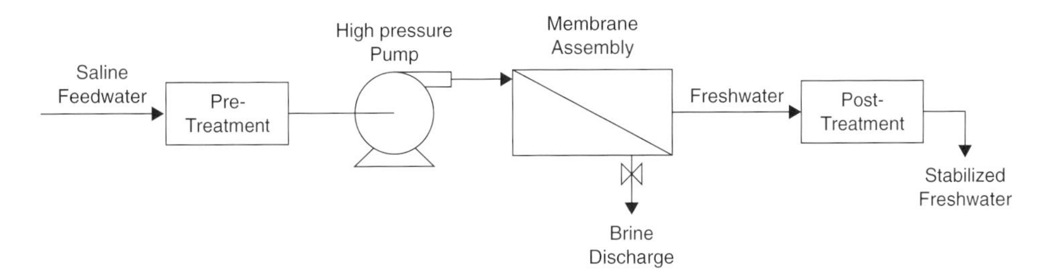

Figure 3.11 Flow diagram of a reverse osmosis system

Source: Courtesy of USAID Khan, 1986

3.5 WATER TRANSFERS

Water transfers are a common component of many urban water supply systems particularly in arid and semi-arid regions. They are not only used to satisfy increasing water demands but also for managing the impacts of droughts. There are many forms of water transfers that serve a number of different purposes in the planning and operation of urban water systems. Eheart and Lund (1996) define the major types as: permanent transfers; contingent transfers/dry-year options (long term, intermediate term, and short term); spot market transfers; water banks; transfer of reclaimed, conserved and surplus water; and water wheeling or water exchanges (operational wheeling, wheeling to store water, trading water of different qualities, season wheeling and wheeling to meet environmental constraints). The major benefits and uses of transferred water include: directly meeting demand and reducing costs; improving system reliability; improving water quality; and satisfying environmental constraints (Eheart and Lund, 1996). The implementation of water-transfers implies the need to increase integration and cooperation among diverse water users.

In arid and semi-arid regions, large water-transfer projects have been used as one solution to meet urban water demand because of the regional differences in water availability. Two very large projects are the Great Man-Made River Project (GMMRP) in Libya; and the newly planned south to north water transfer in China, in which nearly 45 billion cubic metres of water from the Yellow, Yangtze and other rivers will be sent north each year, once the project is finished in 2050.

The GMMRP is a water conveyance system that, when completed, will transport water from aquifers in the south of Libya to the north of Libya. Approximately 80% of this water will be used for agricultural irrigation. Because of the extensive pumping of groundwater in the north along the coast, a substantial seawater encroachment of the coastal groundwater system has occurred. The GMMRP will pump groundwater from the aquifers of Sarir and Kufra in the south-east and Murzag in the south-west of Libya. The project began in 1983 and comprises several phases, Phase I (Sarir/Sirt Tazerbo/Benghazi System) and II (Hasouna/Jefara System), serving the west of Libya, being the two major components.

Another transfer project in arid regions is the National Water Carrier System in Israel, which is supplied mainly by Lake Kinneret. About 80% of the water used in Israel comes from three principal sources: Lake Kinneret, the Coastal Aquifer and the Western Mountain Aquifer.

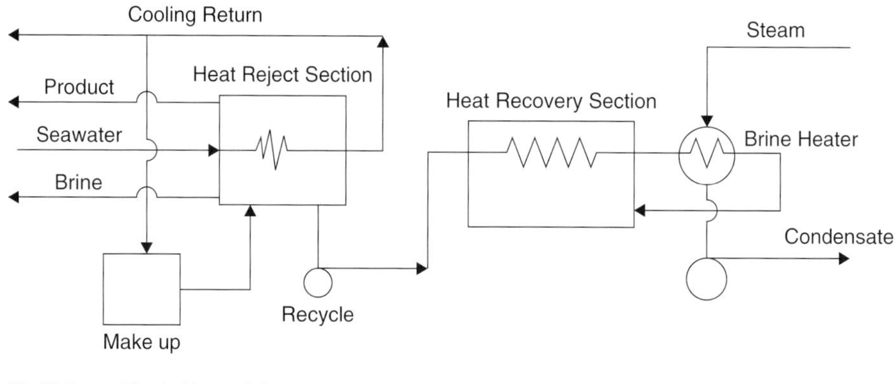

Multi-State Flash (Recycle)

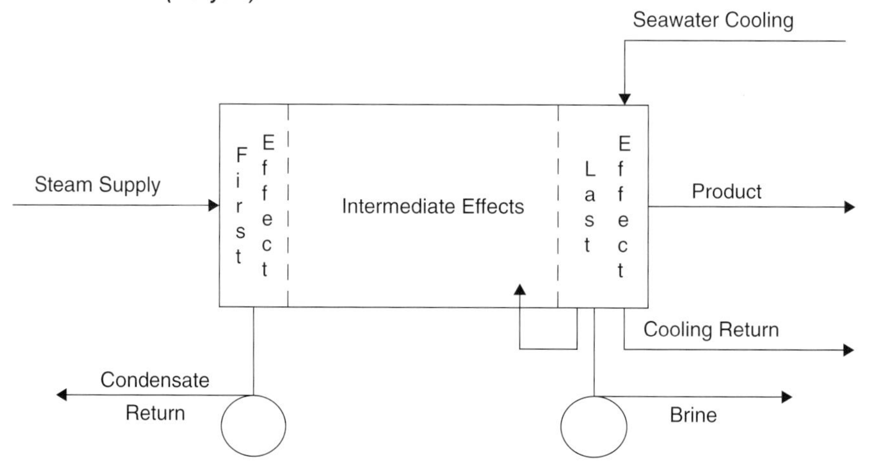

Multiple Effect

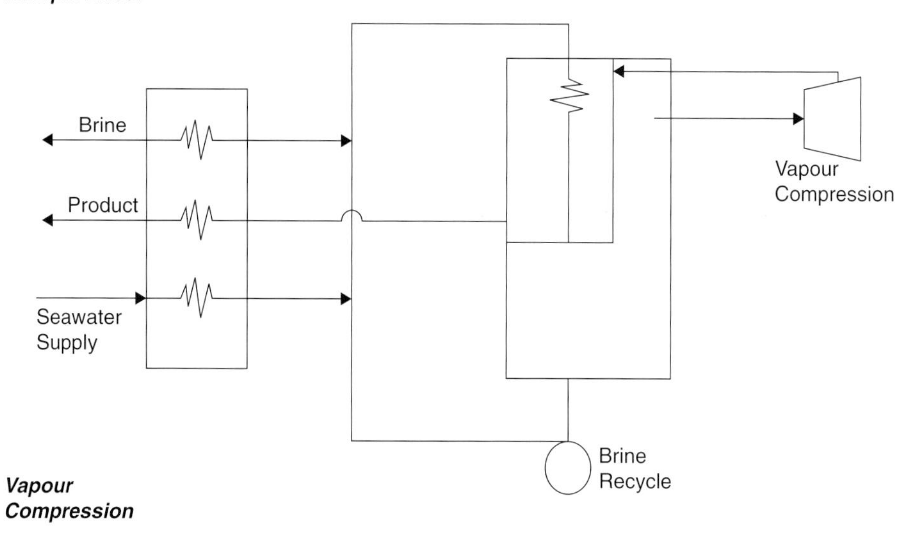

Vapour Compression

Figure 3.12 Common methods of distillation

Source: California Coastal Commission, Seawater Desalination in California, Pantell, 1993

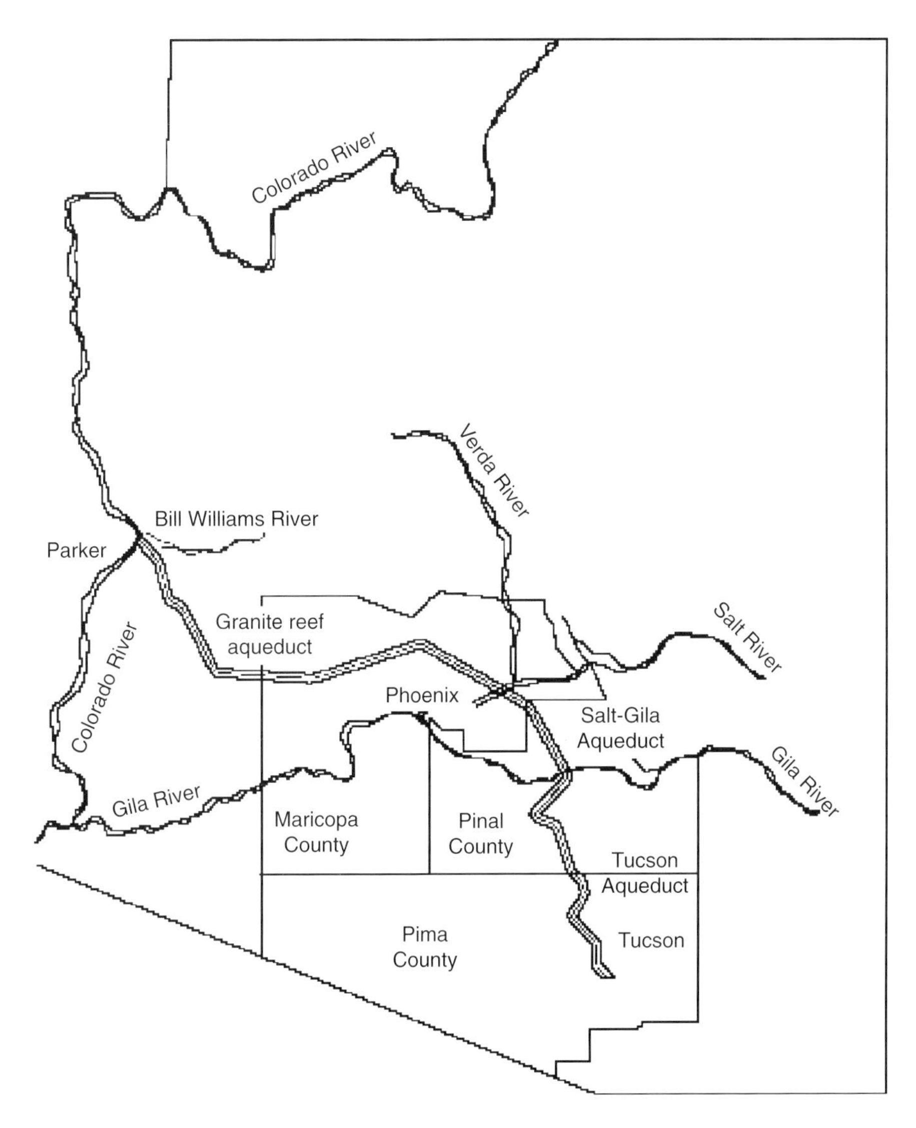

Figure 3.13 Central Arizona Project Canal

Source: Courtesy of CAP

The Central Arizona Project (CAP) conveys water from the Colorado River to central and southern Arizona (see Figures 3.13 and 3.14). This project is a 336-mile long system of aqueducts, tunnels, pumping plants and pipelines. The CAP is designed to bring 1.415 million acre-ft (MAF) of Arizona's 2.8 MAF Colorado River allocation. Of the Lower Colorado River allocation, the CAP has the lowest priority and must curtail its usage in shortage years. This has given a false sense of security to many.

CAP was the largest, most expensive and most politically volatile water-development project in US history. The cost of CAP water and of the associated delivery system made the water cost-prohibitive for agriculture, for which the system was originally conceived, and so it is sold to the public. Both the recharge of CAP and recovery from the aquifer (Tucson, Arizona) and the direct delivery for municipal use are used. Nearly the entire flow of the Colorado River is diverted to Southern California and Arizona. Because of these diversions, today the Colorado River delta in northern Mexico is 'a desiccated place of mud-cracked earth, salt flats, and mushy pools', as noted by Postel (1977).

A much smaller example of a planned future water transfer is the plan for Prescott, Arizona. This proposed transfer is the plan for sustainability that will transfer groundwater from an adjacent groundwater basin called the Big Chino. This effort is to support the increase in population from the present population of 111,000 people to a projected population of 182,000 people by 2025. In other words, the predominant goal or objective is growth. As noted out in the case study, 'it is difficult to imagine a sustainable condition existing beyond year 2025 that does not include the now-taboo provision of growth control'. The environmental integrity of this plan is unknown.

Figure 3.14 Central Arizona Project aqueduct through residential area in Scottsdale, Arizona (See also colour plate 6)

Source: US Bureau of Reclamation

Mexico City is another example of the use of water transfers, which is described in more detail in the case study. Groundwater (5 m^3/s) is transferred from an aquifer located a distance of 100 km and at an elevation of 300 m above Mexico City. Surface water (15 m^3/s) is transferred from the Cutzamala River, located at a distance of 130 km and 1,100 m below Mexico City. The consequences of these water transfers are discussed in the case study. The transfer from the Cutzamala Region reduced the water available for power generation and caused a large loss of irrigation area. Transfers from

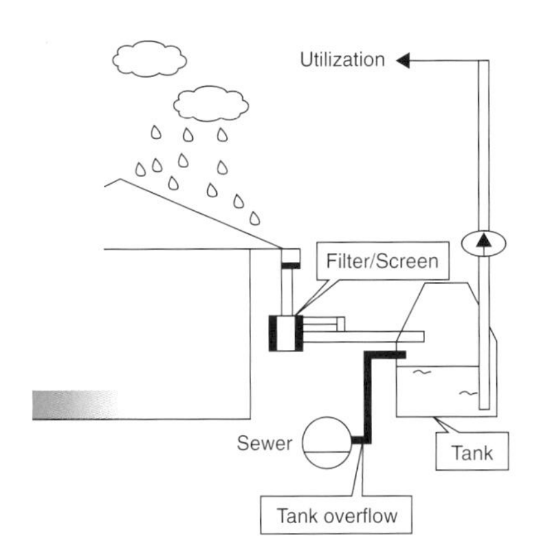

(a) Total flow type rainwater harvesting system. Runoff is confined to a storage tank after passing through a filter or screen before the tank. Overflow to the drainage system only occurs when storage tank is full.

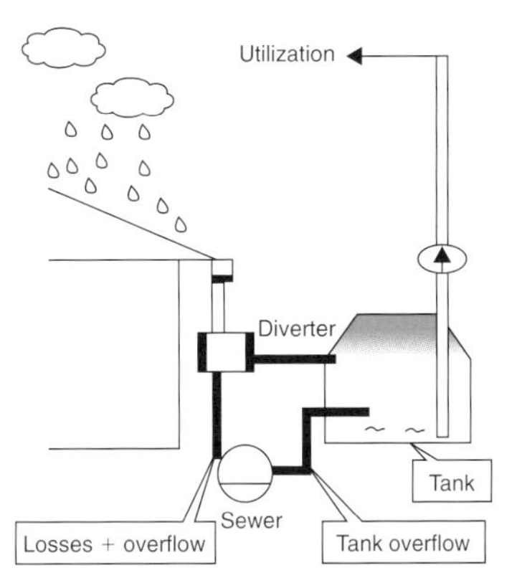

(b) Diverter type rainwater harvesting system. This type contains a branch installed in the vertical rainwater pipe either after the gutter or in the underground drainage pipe. Collected portion is separated from the gutter at the branch and the surplus is diverted to the sewer system. Branches contain fine-meshed sieves that diverts most of the particles to the sewer.

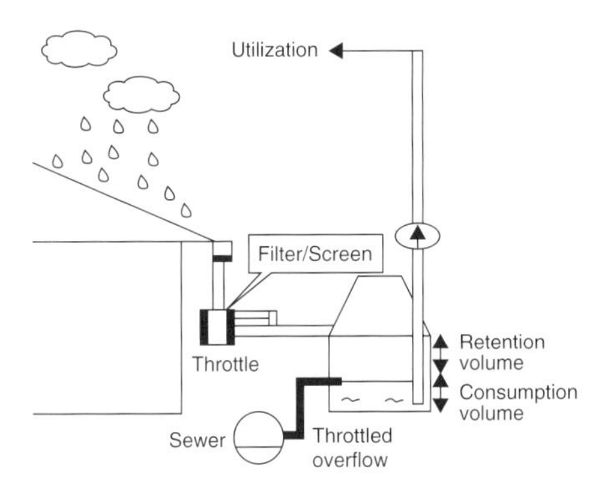

(c) Retention and throttle type rainwater harvesting system. Storage tank provides an additional retention volume, which is emptied using the throttle to the sewer system.

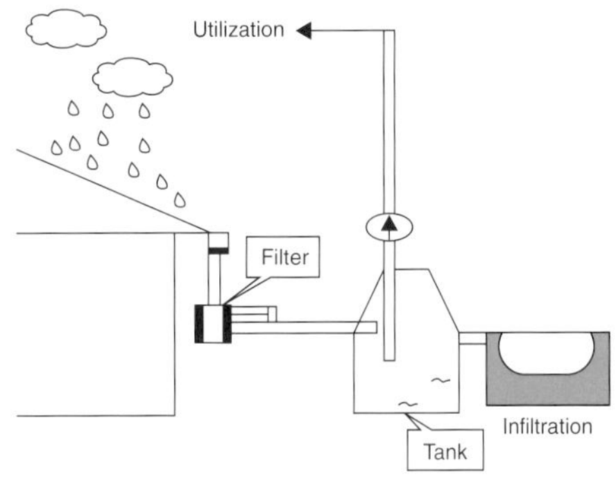

(d) Infiltration type rainfall harvesting system. Local infiltration of the surplus tank overflow is possible instead of diversion to the sewerage system.

Figure 3.15 Types of rainwater harvesting systems

Source: Nouh and Al-Shamsy, 2001; UNESCO IHP-V, Technical Documents in Hydrology, No. 40 Vol.III, UNESCO, Paris

the Lerma region, which used to be a lake, saw the disappearance of fishing and a decrease in water surface elevation in Chapala Lake.

3.6 RAINFALL HARVESTING

From the early civilizations, people in arid and semi-arid regions have relied on collecting or 'harvesting' surface water from rainfalls and storing the water in human-made reservoirs or 'cisterns'. Not only were cisterns used to store rainfall runoff they were also used to store aqueduct water. During ancient times, cisterns ranged from irregular shaped holes (tanks) dug out of sand and loose rock and then lined with plaster (stucco) to waterproof them, to the construction of rather sophisticated structures such as the ones built by the Romans (i.e. one of the largest, the Piscina Mirabillis in Bacoli near Naples supplied by the Augustan (Serino) aqueduct).

In the central Negev Desert in Israel, the six urban centres (Avdat, Mamshit, Nizzana, Shivta, Rehovot and Haluza) were developed by the Nabateans in the hills during the Nabatean-Byzantine times (third to first century BC). Sophisticated rainwater-harvesting systems were developed as the urban water supply for the centres with populations ranging from 25,000 to 71,000 people (Broshi, 1980; Bruins, 2002). Cisterns were dug into the rock and conduits were designed to collect the runoff from roofs, pavements and natural catchments.

Rainfall harvesting can be used as a supplementary, or even primary water source, at the household or small community level (Marsalek et al., 2006). Zuhair et al. (1999) have shown that rainwater harvesting can provide a significant amount of freshwater in the Arabian Gulf States. Examples include: Kuwait (12% of the water demand for landscape agriculture); Muscat, Oman (27% of the water demand for industry); Abha, Saudi Arabia (11% of water demand for industry and landscape irrigation); and Ali Ain, UAE (16% of water demand for agriculture). Figure 3.15 (see previous page) shows four types of rainwater harvesting systems which are systematically distinguished according to their hydraulic properties (Nouh and Al-Shamsy, 2001).

Natural sedimentation in the storage tanks of the rainfall harvesting systems is the most effective cleansing process for roof runoff. It is important to avoid turbulent mixing in the storage tank to prevent the sediment mixing in the water column. A sieve of size 0.5 to 1.0 mm is recommended to prevent residues from entering the pump and installation (Nouh and Al-Shamsy, 2001). Other treatment, such as chemical disinfection, is not necessary. Hermann and Schmida (1999) made a long-term simulation of 10 years of precipitation data to identify various hydraulic factors of rainwater harvesting systems.

Chapter 4

Integrated water excess management in arid and semi-arid regions

The status of water excess management as a part of integrated water systems varies from one country to another. This variation depends primarily on the level of development and the society's awareness of the importance of water excess management. In many places, excess water is considered as a resource which can be retained for reuse, recharged to aquifers or used to create a habitat for wildlife.

4.1 OVERALL SUBSYSTEM COMPONENTS AND INTERACTIONS

Water excess management systems, which include both the stormwater management system and the floodplain management system, can also be thought of as consisting of two respective separate systems: the minor system for storm drainage and the major system for emergency flows. The minor drainage system has been referred to as the 'initial' system or the 'convenience' system (Grigg, 1996; Mays, 2001). Minor systems include gutters, small ditches, culverts, storm drains, detention ponds and small channels. Major systems include the streets and urban streams, floodways and flood fringe areas. The water quality subsystem is superimposed on top of the minor and major systems, as problems arising from the wash-off of surface pollutants, from combined sewer overflows, or from the erosion of pollutants from the inside of sewers.

4.1.1 Stormwater management

The overall key component of stormwater management is the drainage system, which has the following key components (Urbonas and Roesner, 1993):

- The removal of stormwater from streets permitting the functioning of transportation arteries.
- The drainage system controls the rate and velocity of runoff along gutters and other surfaces to reduce the hazards to residents and the potential for damage to pavement.
- The drainage system conveys runoff to natural or manmade major driveways.
- The system can be designed to control the mass of pollutants arriving at receiving water.

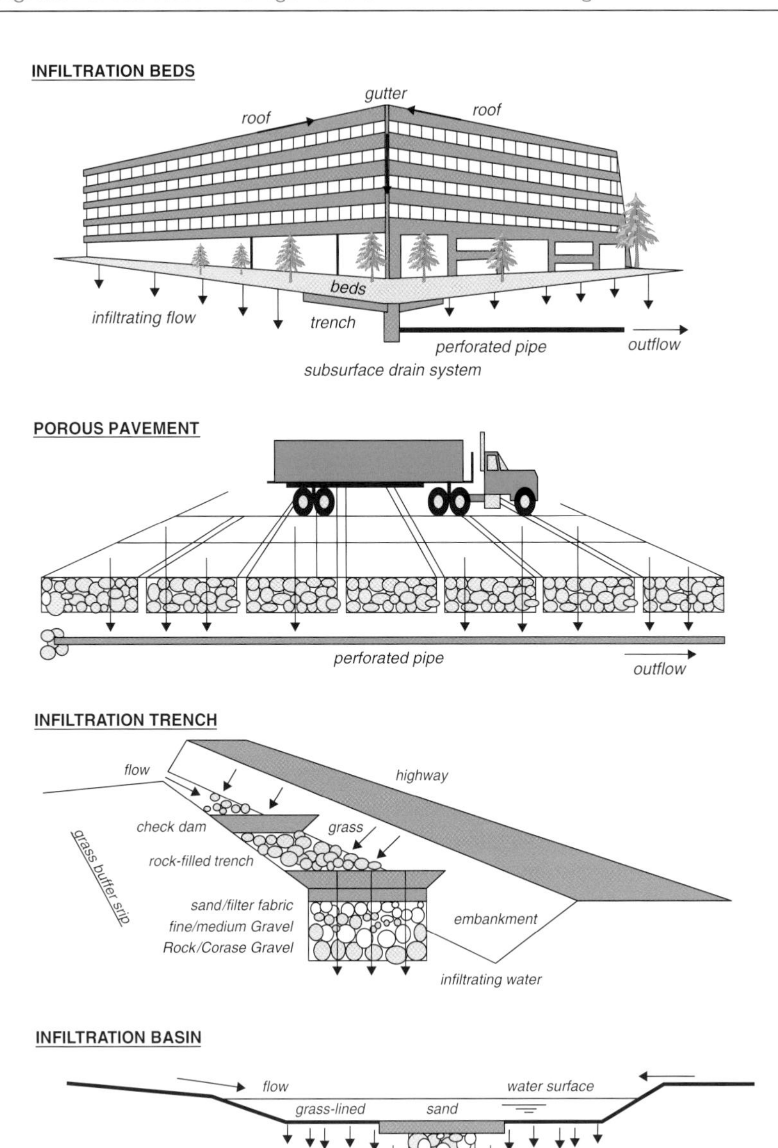

Figure 4.1 Infiltration Devices

Source: Guo, 2001

- Major open drainage ways and detention facilities offer opportunities for multiple use such as recreation, parks and wildlife preserves.

Storm drainage criteria are the foundation for developing stormwater control (Mays, 2001). These criteria should set limits on development; provide guidance and methods of design; provide details of key components of drainage and flood control

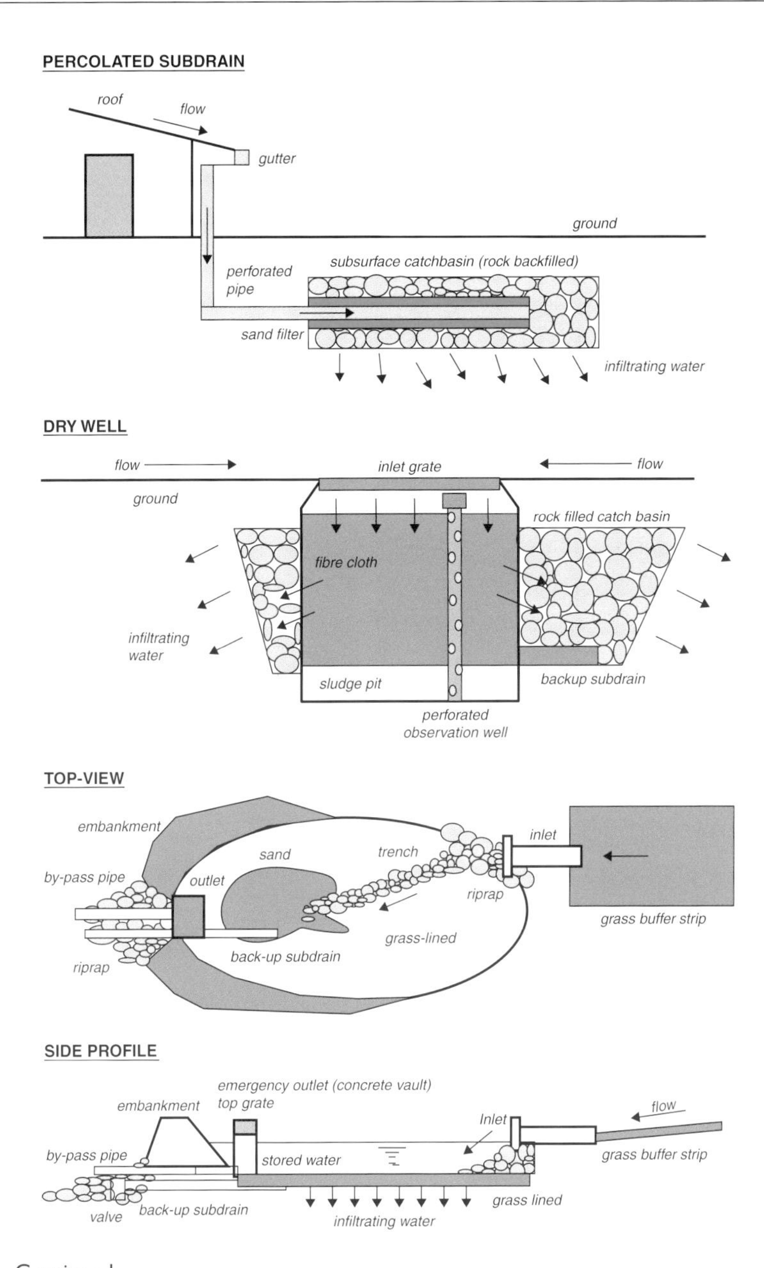

Figure 4.1 Continued

systems; and ensure longevity, safety, aesthetics and maintainability of the system served (Urbonas and Roesner, 1993).

Stormwater management practices to control urban runoff can be classified into seven categories (Mays, 2001). See Figures 4.1–4.4 for illustrations of these practices.

Figure 4.2 Showing rooftop drainage flowing into infiltration bed (photos by L.W. Mays) (See also colour plate 7)

- Infiltration practices;
  - Infiltration basins;
    - Infiltration beds (Figure 4.2);
    - Infiltration trenches;
    - Drywells; and
    - Pervious or porous pavements.
- Vegetated open channel practices (Vegetated open channels are explicitly designed to capture and treat runoff through infiltration, filtration, or temporary storage).
- Filtering practices;
  - Surface sand filter;
  - Underground sand filter;
  - Percolated subdrain;
    - Organic filter;
    - Pocket sand filter; and
    - Bioretention areas.
- Detention ponds (Figures 4.1 and 4.4) or vaults;
  - Extended detention dry ponds (Preferred in arid regions and acceptable in semi-arid regions).
- Retention ponds (Not recommended in arid climates and of limited use in semi-arid regions).

Figure 4.3 Example of small neighbourhood detention basin in Scottsdale, Arizona (photo by L.W. Mays) (See also colour plate 8)

- — Micropool extended detention ponds;
- — Wet ponds;
- — Wet extended detention ponds; and
- — Multiple pond systems.

- Wetlands (Not recommended in arid regions and of limited use in semi-arid regions).
- Other practices such as water quality inlets.
  - — Water quality inlets;
  - — Hydrodynamic devices;
  - — 'Baffle boxes';
  - — Catch basin inserts;
  - — Vegetated filter strips;
  - — Street surface storage;
  - — On-lot storage; and
  - — Microbial disinfection.

Infiltration practices are important in arid and semi-arid regions because they have the following features:

- remove watershed born pollutants
- reduce runoff volume

- provide groundwater recharge
- minimize thermal impacts on fisheries
- augment low flow stream conditions.

4.1.2 Floodplain management

Human settlements and activities have always tended to use floodplains, particularly in urban areas. Their use and occupation has frequently interfered with the natural floodplain processes, causing inconvenience and catastrophe to humans. Floodplains in arid and semi-arid areas include not only the rivers and streams but also alluvial fans. Alluvial fans are characterized by a cone or fan-shaped deposit of boulders, gravel and fine sediments that have been eroded from mountain slopes and transported by flood flows, debris flows, erosion, sediment movement and deposition, and channel migration. The objective of flood control is to reduce or to alleviate the negative consequences of flooding.

According to the US National Flood Insurance Program (NFIP) regulations administered by FEMA, floodplain management is the operation of an overall programme of corrective and preventative measures for reducing flood damage, including, but not limited to, emergency preparedness plans, flood control works and floodplain management regulations. Floodplain management regulations are the most effective method for preventing future flood damage in developing countries with known flood hazards. Floodplain management investigates problems which have arisen in developed areas and potential problems that can be forecasted due to future developments. The encroachment of floodplains, such as the use of artificial fill, reduces the flood-carrying capacity thereby increasing the flood elevations and increasing the flood hazards beyond the encroachments.

Measures that modify the flood runoff are usually referred to as flood control facilities. They consist of engineering structures or modifications. Structural measures (flood control facilities) included reservoirs, diversions, levees or dikes, and channel modifications. Flood control measures that modify the damage susceptibility of floodplains are referred to as nonstructural measures. They include flood proofing, flood warning and land use controls.

Because of the potential for severe flash-flooding in urban areas, particularly in arid and semi-arid regions, flood early warning systems have been constructed and implemented. Flood early warning systems are real-time event reporting systems that consist of remote gauging sites with equipment to transmit information to a base site. The system collects, transmits and analyzes data, and then makes a flood forecast to maximize the warning time to occupants in the floodplain. Figure 4.5 illustrates the flood early warning system used by the Flood Control District of Maricopa County (FCDMC), Arizona.

Urban flood management in developing countries is affected by: development with little or no planning; high population concentration in small areas; lack of stormwater and sewage facilities; polluted air and water; difficult to maintain water supply with a growing population; and poor public transportation, among other things. The urban poor are often forced to settle in flood-prone areas and they lack the adaptive capacity to cope with flood events. The unplanned urbanization and poverty dramatically increase the vulnerability to floods. The increase in vulnerability of cities to flood disasters arises predominantly from the systematic degradation of natural ecosystems, increased urban migration, unplanned occupation, and unsustainable planning and building.

(a) Intake structure to detention pond

(b) Outlet structure for pond

Figure 4.4 Detention basin in Phoenix, Arizona showing inlet and outlet structures (photo by L.W. Mays) (See also colour plate 9)

4.1.2.1 Reactions to flood disasters

The reactions to flood disasters can be categorized into a cycle of restoration (Kahan et al., 2006) divided into three stages: 1) anticipation of the next possible flooding event; 2) actuality of an event; and 3) the aftermath, which considers both the recuperation from the event and decides what changes are needed to better anticipate the next event. The cycle of restoration may start with the planning, followed by the detection, preparation, first response, reconstruction, compensation and learning lessons. These are ordered steps, but there is some overlap in the stages.

Even though Hurricane Katrina did not occur in an arid or semi-arid region, many lessons were learned (Kahan et al., 2006) that could be applied to arid and semi-arid regions:

- Government officials need to consider policies and plans that are more robust against a wider range of disaster scenarios (e.g., on the Gulf of Mexico coast, storm surges had been anticipated, but not the level of Katrina, even though they had anticipated catastrophic flooding and levee failures).

- Failure to anticipate the widespread regional breakdown in infrastructure and services, and the disabling of first-response and public safety programmes were the biggest blind spots throughout the region. (e.g., the planning for regional infrastructure and services must cover total catastrophic breakdown and must include secondary, contingency responses that can be invoked when primary responses are overwhelmed).
- Detection of the storm was adequate, but the detection of structural weakness, soil anomalies and impending failure was not, as no monitoring was in place. (This has been remedied through extensive deployment of sensors on all structural features of the flood protection system).
- Reconstruction efforts are strongly influenced by the answer to the question, what will the level of protection be in the future? Complicating this is the fact that many flood victims have chosen not to return and economic recovery remains uncertain.
- Integrated urban water management from the perspective of flood control includes conceding land to the water from time to time (somewhat psychologically and politically difficult).

Kahan et al. (2006) outlined several overarching lessons from their study of example flood events and the Hurricane Katrina events:

- Building bigger and better flood protection works does not necessarily maximize safety.
- Differing perceptions among residents and political leaders of the permanence and transience of the physical environments can create conflicts in decisions about what to rebuild, what to modify and what to leave as is.
- Some potential improvements to the status quo ante are not intuitively apparent or politically palatable.
- Structural solutions are necessary but not sufficient.

4.1.3 Stormwater/floodplain management

Stormwater management and floodplain management are generally separate and different programmes; however, there are many situations when these two should be coordinated management efforts. Unfortunately in many locations, such as in the US, these are separate and sometimes conflicting efforts. One example is the placement of detention and retention basins for stormwater management purposes, particularly in urban areas. Treatment plants and other water-related structures are also placed in floodplains in urban environments limiting the flow capacity of the floodplain. Detention basins generally should be located outside the floodplain, particularly for small streams of relatively small drainage areas. In such cases, the same storms affect the development site and the floodplain simultaneously. The time during which the detention basin is needed to store stormwater from the development is the same time that the floodplain is flooded and the detention basin location is already filled with floodwater. To the extent that the flood at the development coincides with the flooding in the floodplain, detention storage in the floodplain is ineffective. The large amount of fill required for the detention basin makes the floodplain less effective and increases flood elevations.

In summary, effective water excess management requires simultaneous stormwater management and floodplain management. To be effective this needs to be a coordinated management effort.

4.1.4 Wadi flash flooding

Wadi flash flooding is extremely important in many arid lands in the Middle East and North Africa (MENA). Because of the short hydrologic response time, it is important to understand flash flooding and provide adequate warning. It is also important to understand the potential for groundwater recharge which provides a valuable renewable water resource. Numerous studies in arid land hydrology have estimated transmission losses to range from 40 to 50% of the flood volume lost in wadis. Almost half of the remaining rain evaporates before it becomes effective rainfall.

Flood forecasting and real-time, flood early warning systems are very valuable tools that need to be implemented to save lives in many areas of the MENA and other arid and semi-arid areas. The components of early flood warning systems for wadis are monitoring of meteorological and hydrologic data, automated processing, real-time transmission of information and wadi network modelling.

A UNESCO and Regional Center on Urban Water Management (RCUWM - Tehran) sponsored an international workshop on Flash Floods in Urban Areas and Risk Management in September 2006 and presented the following comments and recommendations that pertain to wadi flash-flooding:

- Conduct flash-flood studies focusing on objectives, feasibilities, limitations and benefits of flash-flood structural and nonstructural systems.
- Application needs to be made of new technologies including RS/GIS – MIS.
- Build and operate dams/reservoirs at appropriate wadi locations for flood control and artificial recharge of groundwater.
- Use physically based simulation models to understand the atmosphere, surface water for rainfall runoff and groundwater to estimate the quality and quantity of surface and groundwater.
- For flood early warning systems, there should be coordination among government organizations, preventive methods should be checked onsite, and action plans that provide notification and evacuation should be developed.

4.2 IMPACTS OF URBANIZATION ON STORMWATER

Urban stormwater runoff includes all flows discharged from urban land uses into stormwater conveyance systems and receiving waters. Urban runoff includes both dry-weather non-stormwater sources (e.g., runoff from landscape irrigation, dewatering, and water line and hydrant flushing) and wet-weather stormwater runoff. The water quality of urban stormwater runoff can be affected by the transport of sediment and other pollutants into streams, wetlands, lakes, estuarine and marine waters, and groundwater. The costs and impacts of water pollution from urban runoff are significant and can include: fish kills; health concerns of human and/or terrestrial animals; degraded drinking water; reduced water-based recreation and tourism opportunities;

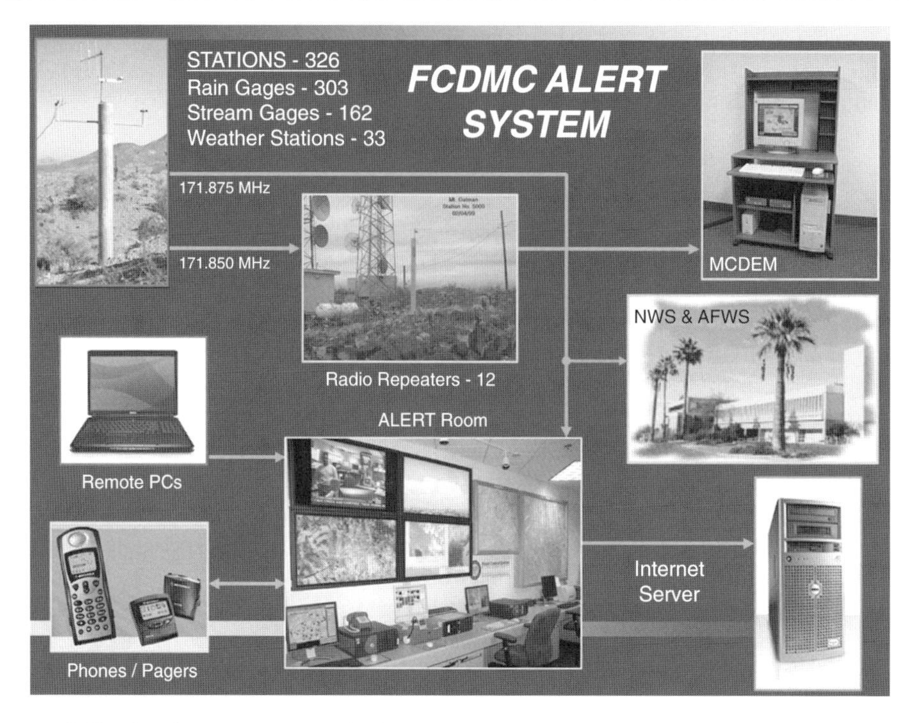

Figure 4.5 Flood early warning system for Flood Control District of Maricopa County (FCDMC) (See also colour plate 10)

Source: Courtesy of FCDMC

economic losses to commercial fishing and aquaculture industries; lowered real estate values; damage to the habitat of fish and other aquatic organisms; inevitable costs of clean-up and pollution reduction; reduced aesthetic values of lakes, streams and coastal areas; and other impacts (Leeds et al., 1993).

Increased stormwater flows from urbanization have the following major impacts (FLOW, 2003):

- acceleration of stream velocities and degradation of stream channels
- declining water quality due to the washing off of accumulated pollutants from impervious surfaces to local waterways, and an increase in siltation and erosion of soils from pervious areas subject to increased runoff
- increase in the volume of runoff, with higher pollutant concentrations that reduce receiving water dilution effects
- reduced groundwater recharge, resulting in decreased dry-weather flows, poorer water quality of streams during low flows, increased stream temperatures and greater annual pollutant load delivery
- increased flooding
- combined and sanitary sewer overflows due to stormwater infiltration and inflow
- damage to stream and aquatic life, resulting from suspended solids accumulation, and increased health risks to humans from trash and debris which can also endanger and destroy food sources or habitats of aquatic life (FLOW, 2003).

4.3 RECOMMENDATIONS FOR RESEARCH

The following are recommendations for further research.

Effects of sediments in sewers

- Sediment deposits in sewers create negative effects on the hydraulic performance of the system and on the environment.
- Problems that arise include blockage, surcharge, early overflows, large pollutant discharges and costly removal.
- Very little research has been done in this area.

Research needs for water excess management in ASA

- Advancement of alert/warning systems.
- Stormwater/floodplain management in alluvial fan areas.
- Erosion/sediment in stormwater management systems.
- Risk/reliability approaches and optimization to design and analyze stormwater management.
- Better coordination of stormwater management and floodplain management.

Chapter 5

Interactions and issues of urban water management

5.1 PRINCIPLES OF INTEGRATED WATER RESOURCES MANAGEMENT (IWRM)

Integrated water management can be considered in at least three ways (Mitchell, 1990), including:

1. The systematic consideration of the various dimensions of water: surface and groundwater, quality and quantity.
2. The implication that while water is a system it is also a component which interacts with other systems.
3. The interrelationships between water and social and economic development.

In the first thought, the concern is the acceptance that water comprises an ecological system, which is formed by a number of interdependent components. In the second one, the interactions between water, land and the environment, which involve both terrestrial and aquatic issues, are addressed. Finally, the concern is with the relationships between water and social and economic development, since availability or lack of water may be viewed as an opportunity for or a barrier against economic development. Each aspect of integrated management depends on and is affected by other aspects.

The principles of IWRM are summarized in Table 5.1.

5.2 WATER LAWS AND POLICIES

5.2.1 South-western United States

Water management decisions are most often underlain by water laws. In the United States, water law has two basic functions: 1) the creation of supplemental private property rights in scarce resources, and 2) the imposition of public interest limitation on private use. For our purposes, water law is divided into surface water law and groundwater law. Surface water law is further categorized into riparian law and appropriation law. Riparian law is based on the riparian doctrine which states that the right to use water is considered real property, but the water itself is not the property of the

Table 5.1 Summary of the principles of integrated water resources management (IWRM) in urban areas (Industry Sector Report for WSSD prepared by IWA http://www.grdc.org/uem/water/iwrm/1pager-01.html)

- — IWRM should be applied at the catchment level.
- — It is critical to integrate water and environmental management.
- — A systems approach should be used.
- — Full participation by all stakeholders, including workers and the community.
- — Attention to social dimensions.
- — Capacity building.
- — Availability of information and the capacity to use it to make policy and predict responses.
- — Full-cost pricing complemented by targeted subsidies.
- — Central government support through the creation and maintenance of an enabling environment.
- — Adoption of the best existing technologies and practices.
- — Reliable and sustained financing.
- — Equitable allocation of water resources.
- — The recognition of water as an economic good.
- — Strengthening the role of women in water management.

landowner (Wehmhoefer, 1989). Appropriation law states that the allocation of water rests on the proposition that the beneficial use of water is the basis, measure and limit of the appropriative right, the first in time is prior in right. In the western United States, surface water policy generally follows this doctrine of 'first in time, first in right'. To appropriate water, the user need only demonstrate the availability of water in the source of supply, show intent to put the water to beneficial use and give priority to more senior permit holders during times of shortage (Schmandt et al., 1988). Beneficial use of water under the law includes: domestic consumption, livestock watering, irrigation, mining, power generation, municipal use and others. The states of Arizona and New Mexico follow the appropriation law of surface waters, and in California and Texas the appropriation doctrine and the riparian doctrine coexist.

Groundwater allocation is handled quite differently and is typically divided into common law or statutory law. Common law doctrines include the overlaying rights doctrines of absolute ownership, reasonable use and correlative rights. These doctrines give equal rights to all landowners overlaying an aquifer. Arizona, California and Texas have adopted these principles for groundwater allocation.

The above surface and groundwater laws serve as the basis for individual state water policies. The burden of developing water policies lies with the states. This is often achieved by the state proposing a water project and securing federal funds for the construction. It is also up to the states to agree on apportionment in interstate waters, if the states cannot agree, then the courts will intervene and settle the dispute by decree. The federal government only becomes involved in such disputes where federal lands and Indian reservations are concerned.

5.2.1.1 Arizona

It is no secret that throughout Arizona's history, water policy has been directed at supporting the unconstrained growth of its population and major revenue-producing activities. Starting with mining, ranching and farming, with the gradual shift to municipal and industrial uses, the water policy of the state has been directed at obtaining imported supplies. This has been an effort to augment what has appeared to be an insufficient indigenous resource. Waterstone (1992) points out that the

> state's water policies has led to the protracted exercise to capture and secure the Central Arizona Project (CAP), the ongoing infatuation with weather and water shed manipulation, the current experimentation with groundwater recharge and effluent use, and the recent spate of purchases of remote water farms'.

In Arizona, the state's water policy and management focused more on surface water than groundwater prior to 1980, when the Groundwater Management Code (GMC) was developed; thereafter, the emphasis has been on groundwater. In regards to surface water, Arizona law defines surface water as 'the waters of all sources, flowing in streams, canyons, ravines or other natural channels, or in definite underground channels, whether perennial or intermittent, flood, waste, or surplus water, and of lakes, ponds and springs on the surface'. These surface waters are subject to the 'doctrine of prior appropriation' (ADWR, 1998). In Arizona, surface water rights are obtained by filing an application with the Department of Water Resources for a permit to appropriate surface water. Once the permit is issued and the water is actually put to beneficial use, proof of that use is made to the Department and a certificate of Water Right is issued to the applicant. Once a certificate is issued, the use of the water is subject to all prior appropriations.

Since water law in the state of Arizona has changed substantially over the years, Arizona is now conducting a general adjudication of water rights in certain parts of the state. Adjudications are court determinations of the status of all State law rights to surface water and all claims based upon federal law within the river systems. These adjudications will provide a comprehensive way to identify and rank the rights to the use of water in some areas. The adjudications will also quantify the water rights of the federal government and the Indian reservations within Arizona.

In Arizona, groundwater problems arise from the overdrafting of water from the aquifers. Groundwater overdrafts cause many problems, such as increased well pumping costs and water quality issues. In areas of severe groundwater depletion, the earth's surface may also subside, causing cracks or fissures that can damage roads or building foundations. To manage groundwater pumping in Arizona, the Arizona Groundwater Management Code (GMC) was developed in 1980 as state legislation. The Arizona GMC was named one of the nation's ten most innovative programmes in state and local government by the Ford Foundation in 1986. This achievement came from the cooperation of Arizonans working together and compromising when necessary to protect the future of the state's water supply.

The Groundwater Management Code has three primary goals (ADWR, 1998):

1. Control the severe overdraft currently occurring in many parts of the state.
2. Provide a means to allocate the state's limited groundwater resources to most effectively meet the changing needs of the state.
3. Augment Arizona's groundwater through water supply development.

To achieve these goals, the code set up a comprehensive management environment and established the Arizona Department of Water Resources.

The code outlines three levels of water management. Each level is based on different groundwater conditions. The lowest level applies statewide, and includes general groundwater provisions. The next level applies to Irrigation Non-Expansion Areas (INAs); and the highest level applies to Active Management Areas (AMAs) where groundwater depletion is the highest. The boundaries that divide the INAs and AMAs are determined by groundwater basins and not by political jurisdiction. The main purpose of groundwater management is to determine who may pump groundwater and how much may be pumped. This includes identifying existing water rights and providing new ways for non-irrigation water users to initiate new withdrawals. In an AMA or INA, new irrigation users are not allowed. Even with the original publicity and enthusiasm, many people now feel that the efforts under the Groundwater Management Code have been very costly with very little savings in water, making the success questionable.

5.2.1.2 Texas

In the state of Texas, one way to accomplish the wide range of water management duties has been the development of river authorities. The term 'river authority' implies an institution that possesses authority over a river, thereby imparting a regional character to the organization (Harper and Griffin, 1988). Some river authority boundaries are defined by watershed boundaries and some by county boundaries. Some of these river authorities share jurisdiction over an entire watershed and only seven out of the thirteen are the sole river authority operating in their particular basin. The duties and powers of the river authorities can be divided into the following groups: 1) watershed management, 2) water supply, 3) pollution control and groundwater management, 4) appurtenant development and 5) governmental or administrative authority.

The right of landowners to intercept and use diffused surface water on their properties is superior to that of adjacent landowners and to any holder of surface water rights on streams into which the runoff might eventually flow. Diffused surface water is drainage over the land surface before it becomes concentrated in a stream course. Concerning stream flow, Texas is a dual-doctrine state, recognizing both the riparian doctrine and the prior appropriation doctrine. Riparian doctrine is a complex blend of Hispanic civil law and English common law principles. The prior appropriation doctrine was adopted before the turn of the century for allocating surface-water rights. This dual doctrine has caused great difficulty in coordinating the diverse private and public water rights emanating from diametrically different doctrines. Surface water rights adjudication began in 1969 to merge all unrecorded surface-water rights into a permit system. This has simplified the complex management issue. In summary, private or landowner rights pertain to percolating groundwater and diffused surface-water, and the state has appropriated the flow of rivers and streams.

The state of Texas is large and diverse, which means that water management solutions that are appropriate for one region may not be appropriate for another. To implement the water policy in Texas, the state has developed a comprehensive water plan which guides surface and groundwater management. The common element underlying the State of Texas's water planning is the fact that meeting the future water needs of the state will require a full range of management tools. These management tools are listed below. Water availability, economics, environmental concerns and public acceptance identify which management tool is best for a specific water need.

- Expected water conservation
- Advanced water conservation
- Water reuse
- Expanded use of existing supplies
- Reallocation of reservoir storage
- Water marketing
- Subordination of water rights
- Chloride control measures
- Interbasin transfers
- New supply development

Surface water policies in Texas are governed by both the prior appropriation doctrine and the riparian doctrine, and are integrated into the water plan. Surface water in Texas is held in public trust by the state and is allocated to users through a system of water rights. The Texas Water plan delegates most of the planning authority to river authorities and water districts. Due to this type of delegation, conflicts can arise between the various water planning entities operating in the same river basin.

A large amount of Texas's water supply is from groundwater resources, which has resulted in a severe depletion of some of the major aquifers in the state. It is surprising that a state such as Texas, that depends on groundwater so much, has such an ill-defined groundwater plan. Texas groundwater law is based on the 'absolute ownership rule', which states that 'percolating waters are the private property of the landowner' (Schmandt et al., 1988). This means that a landowner has the right to pump water from a groundwater deposit beneath his land at any rate, as long as the withdrawal does not maliciously harm his neighbour. The groundwater law excludes underground streams, where the classification of an underground stream is still a matter of debate. This lack of regulation has caused many problems in the state of Texas such as overdrafting, subsidence and water quality issues. The state of Texas has conducted efforts to develop best management techniques for regional groundwater planning.

5.2.1.3 New Mexico

Similar to the state of Texas, New Mexico also allocates water rights under the legal doctrine of prior appropriation. The state is considered the owner of surface water and it holds it in trust for the public. The water appropriations are made through the state engineer, who administers the water law. All allocated waters are subject to appropriation, with the exception of wells for domestic use, which are defined as wells that have a draw of less than 1800 gpd. All water rights are lost if the water is not used in four consecutive years. To settle water controversies, eight Interstate Stream Compacts were developed.

US Supreme Court decrees govern the use of water in the Pecos and Gila River Basins in New Mexico. Unlike the state of Texas, that governs the use of groundwater by absolute ownership, New Mexico controls the use of groundwater under a system of permits, which uses the priority concept. Since most groundwater aquifers are related hydrologically to surface water, it is also a concern of the state to regulate groundwater pumping. The state requires that applicants for groundwater wells withdraw surface rights to offset the impacts of the pumping. This type of management coordination between surface and groundwater was developed to protect senior water rights.

5.3 INSTITUTIONAL FRAMEWORK

In Chapter 1 the concept of an urban water system implying a single urban water system with the reality that it is an integrated whole was discussed. It was also noted that the concept of a single 'urban water system' is not fully accepted because of the lack of integration of the various components that make up the total urban water system. It should be no surprise then that, historically, institutional frameworks have not been set up to manage such a system.

The challenge is that the role of government and even the form of government are being greatly altered in several countries, while the approach to resource management is undergoing technological and philosophical changes (Frederiksen, 1997). Today, conditions in many parts of the world, including Asia, Africa and the Middle East, have degenerated to the point where countries encounter water shortages under normal precipitation conditions. Countries lack the plans and infrastructure needed to deal with the calamity which will result during the next widespread drought. The sad thing is that in many cases the affected public is unaware of the impending disaster. The institutional frameworks are simply not in place to deal with the situations.

Institutions for the development and management of water resources have evolved over the centuries. Many countries lack water rights systems to record allocations that protect the investors and the public users. In many cases, water quality considerations are not part of the criteria that govern the use of water allocations. Few effective allocation mechanisms are in place in developing countries, other than what the government does or does not construct (Frederiksen, 1997).

5.4 VULNERABILITY OF URBAN WATER SYSTEMS

5.4.1 Natural disasters

Natural disasters that effect urban water management in arid and semi-arid regions include droughts and floods.

Droughts can be classified into meteorological droughts, which refer to lack of precipitation; agricultural droughts, which refer to lack of soil moisture; and hydrological droughts, which refer to reduced stream flow and/or groundwater levels.

5.4.2 Climate change

The climate system is an interactive system consisting of five major components – the atmosphere, the hydrosphere, the cryosphere, the land surface and the biosphere – which is forced or influenced by various external forcing mechanisms, the most important of which is the sun. The effect of human activities on the climate system is considered as external forcing. Climate change predictions are based on computer simulations using general circulation models (GCMs) of the atmosphere. The limitations of state-of-the-art climate models are the primary sources of uncertainty in the experiments that study the hydrologic and water resources impact of climate change. Future improvement to the climate models, hopefully resulting in more accurate regional predictions, should greatly improve the types of experiments to more accurately define the hydrologic and hydraulic impacts of climate change.

Future precipitation and temperature are the primary drivers for determining future hydrologic response. Because of the uncertainties of the predictions of the future precipitation and temperatures, the hydrologic responses of various river basins are uncertain, resulting in uncertainties of our future urban water resources, particularly in arid and semi-arid regions. In general, the hydrologic effects are likely to influence water-storage patterns throughout the hydrologic cycle and affect the exchange between aquifers, streams, rivers and lakes. In arid and semi-arid regions, relatively modest changes in precipitation can have proportionately larger impacts on runoff, and higher temperatures result in higher evaporation rates, reduced stream flows and increased frequency of droughts (Mays, 2007). The effects of climate change on groundwater sustainability include (Alley et al., 1999):

- Changes in groundwater recharge resulting from changes in average precipitation and temperature or in seasonal distribution of precipitation.
- More severe and longer droughts.
- Changes in evapotranspiration resulting from changes in vegetation.
- Possible increased demands for groundwater as a backup source of water supply.

In the Fourth World Water Forum in March 2006, the Cooperative Programme on Water and Climate (CPWC) pointed to 'the alarming gap between international recognition of the risks posed by climate change and the general failure to incorporate measures to combat those risks into national and international planning strategies' (McCann, 2006). The CPWC findings and recommendations to cope with climate extremes were encompassed within the following five key messages:

- Strategies for achieving the 2015 Millennium Development Goals (MDG) do not account for the climate variability and change.
- Climate-related risks are not sufficiently considered in water-sector development and management plans.
- Investment in climate disaster risk reduction is essential.
- The trend of increasing costs has to be reversed through the safety chain concept (prevention, preparation, intervention, risk spreading, recondition, reconstitution).
- Coping measures need to combine a suite of technical/structural and nonstructural measures.

The first message addresses the fact that climate impacts on hydrological systems and on livelihoods threaten to undo decades of development efforts. The second message relates to the fact that to meet MDG targets there needs be substantial long-term investments in structural and nonstructural approaches to water management. Structural measures include storage, control and conveyance; and nonstructural measures include demand-side management, floodplain management, service delivery, etc. The third message relates to the fact that the costs of disasters, especially those related to water, are increasing, and substantial efforts are needed in mainstream climate reduction. The fifth message advocates a combination of both structural and nonstructural measures. Structural measures include dams, dikes and reservoirs; and

Table 5.2 Supply-side and demand-side adaptive options for urban water supply (IPCC, 2001)

Supply Side		*Demand Side*	
Option	*Comments*	*Option*	*Comments*
Increase reservoir capacity	Expensive; potential environmental impact	Incentives to use less (e.g., through pricing)	Possibly limited opportunity, needs institutional framework
Extract more from rivers or groundwater	Potential environmental impact	Legally enforceable water standards (e.g., for appliances	Potential political impact; usually cost-inefficient
Alter system operating rules	Possibly limited opportunity	Increase use of gray water turbines; encourage energy efficiency	Potentially expensive
Interbasin transfer	Expensive; potential environmental impact; may not be feasible		
Desalination	Expensive; potential environmental impact	Reduce leakage	Potentially expensive to reduce to very low levels, especially in old systems
		Development of non-water-based sanitation systems	Possibly too technically advanced for wide application

nonstructural measures include early flood warning systems, spatial planning, 'living with water' insurance, etc.

The reality is that climate change impacts are already with us and are manifesting as increasing occurrences of and intensity of climate extremes, such as droughts, floods and climate variability. From the water supply perspective, there are both supply-side and demand-side options that could be considered for urban water supply. Table 5.2 summarizes some supply-side and demand-side adaptive options for the urban water-use sector.

5.4.3 Human-induced disasters

There are many human-induced events that could affect our urban water supplies (threats to the drinking-water system) ranging from depletion/over-exploitation of groundwater supplies (Todd and Mays, 2005) to possible terrorist activities (Mays, 2004c) to human-induced effects of climate change (Mays, 2007).

Some of the symptoms and susceptibilities of depletion/over-exploitation of groundwater supplies include: declining water levels, degradation of water quality due to seawater intrusion or the induced up-coning of poor quality water, and land subsidence. Declining water levels decrease the natural flows to springs and rivers, which in turn can have ecological impacts, such as the increased risk of land subsidence, and an increase in infrastructure costs for deeper wells and the required pumps and increased energy costs.

Land subsidence accompanies the lowering of the piezometric surface in regions of heavy pumping from a confined aquifer (Todd and Mays, 2005). Land subsidence is a global problem that is usually caused by over-exploitation of groundwater. More than 80% of the identified subsidence in the United States is a consequence of exploitation of groundwater. The semi-arid areas region of the south-western US where subsidence is significant include: Las Vegas, Nevada; the Albuquerque Basin and Mimbres Basin in New Mexico; and in South Central Arizona (Tucson and Phoenix and its surrounding area). As pointed out in the case study on Mexico City, over-exploitation of local groundwater has resulted in:

- a loss of the sewerage/drainage capacity
- serious structural problems with buildings
- leaks in water and wastewater networks
- deterioration of groundwater quality
- metro rails need to be levelled each year and in some areas the accumulated changes are compromising its operation.

The events of September 11, 2001 have significantly changed the approach to the management of water utilities, particularly in the United States. Previously, the consideration of the terrorist threat to the US drinking-water supply was minimal. Now in the US there has been an intensified approach to the consideration of a terrorist threat. The probability of a terrorist threat to drinking water is probably very low; however, the consequences could be extremely severe for exposed populations. Some types of threat may have higher probabilities than others. Some of the threats include cyber threats, physical threats, chemical threats and biological threats. Water storage and distribution systems can facilitate the delivery of an effective dose of toxicant to a potentially very large population. These systems can also facilitate a lower level of chronic dose (for chemicals) with longer-term effects and lower-detection thresholds.

5.4.4 Risk assessment

Any sustainable development of urban water resource projects requires consideration of several goals and involves different stakeholders. Decision making is multi-dimensional in nature which generally involves objectives and constraints arising from political, environmental, social, economical and engineering aspects. Engineering design is often at the final stage to find technical means to best accomplish the project goals. Tung (2007) presents a risk-based analysis framework for sustainable development of water resource engineering projects that integrates uncertainties, risks and multi-criteria decision. Also see Tung and Yen (2005) and Tung et al. (2006) for a more thorough development of methodologies for uncertainty analysis and reliability approaches.

5.5 TOOLS AND MODELS FOR INTEGRATED URBAN WATER MANAGEMENT

Many management tools and models have been developed for use in integrated urban water management. These tools include not only the various simulation models, but also the Decisions Support System (DSS), Supervisory Control Automated Data Acquisition (SCADA) systems, adaptive management and optimization approaches, that include simulation for design operation and management of urban water system.

5.5.1 Decision support systems

Decision support systems, as might be inferred from the name, do not refer to a specific area of speciality. It is not easy to give a specific definition to DSSs based on their uses. Although some consensus exists as to the purpose of DSSs, 'a single, clear, and unambiguous definition is lacking'. Generally, however, DSSs give pieces of information, sometimes real-time information, that help make better decisions. A DSS is an interactive computer-based support system that helps decision-makers utilize data and computer programs to solve unstructured problems. A DSS generally consists of three main components: 1) state representation, 2) state transition, and 3) plan evaluation (Reitsma et al., 1996). State representation consists of information about the system, in such forms as databases. State transition takes place through modelling, such as simulation. Plan evaluation consists of evaluation tools, such as multi-criteria evaluation, visualization and status checking. These three components comprise the database management subsystem, model base management subsystem and dialogue generation and management subsystem, respectively.

Numerous DSSs have been developed over the years, many of which are commercially available packages that include extremely sophisticated components; however, they are deficient in the integration of various interrelations among the different social, environmental, economic and technological dimensions of water resources (Todini et al., 2006). One model of interest is the Water Strategy Man Decision Support System (WSM DSS) which was developed by the EU-funded Water Strategy Man project (Developing Strategies for Regulating and Managing Water Resources and Demand in Water Deficient Regions). According to Todini et al. (2006), this model was developed as a GIS-based package with the objective to emphasize the conceptual links between the various components and aspects of water resource system management, instead of merely focusing on the detailed description of the physical systems and related phenomena. Maia (2006) describes an application of the WSM DSS to evaluate the alternative water management scenarios of the Ribeiras do Algarve in the semi-arid region of southern Portugal. This river basin includes an area of 3837 km^2 and 18 municipalities.

5.5.2 Supervisory control automated data acquisition (SCADA)

SCADA is a computer-based system that can control and monitor several hydrosystems operations such as pumping, storage, distribution, wastewater treatment and so on. Many such systems have been developed for different water supply agencies and flood-control agencies. In general, SCADA systems for water distribution systems are designed to perform the following functions:

- Acquire data from remote pump stations and reservoirs and send supervisory controls.
- Allow operators to monitor and control water systems from computer-controlled consoles at one central location.
- Provide various types of displays of water system data using symbolic, bar graph and trend formats.
- Collect and tabulate data and generate reports.
- Run water-control software to reduce electrical power costs.

Remote terminal units (RTUs) are used to process data from remote sensors at pump stations and reservoirs. The processed data are transmitted to the SCADA system also by the RTUs. Conversely, supervisory control commands from the SCADA system prompt the RTUs to turn pumps on and off, and open and close valves.

5.5.3 Adaptive management

Adaptive management has emerged as an approach to resolve natural resource conflicts in the face of significant uncertainty. Adaptive management is a systematic and rigorous, scientifically defensible programme of learning from the outcomes of management actions, accommodating change and improving management (Walters and Holling, 1990). Sit and Taylor (1998) stated the following:

> Adaptive management is a systematic process for continually improving management policies and practices by learning from the outcomes of operational programs. Its most effective form – 'active' adaptive management – employs management programs that are designed to experimentally compare selected policies or practices, by evaluating alternative hypotheses about the system being managed.

It is further stated that 'active adaptive management can involve deliberate "probing" of the system to identify thresholds in response and to clarify the shape of the functional relationship between actions and response variables.'

Adaptive management is generally described as a recursive process consisting of four main steps: plan, act, monitor and evaluate. Skaggs et al. (2002) discussed the application of adaptive management to water supply planning for Mexico City. They noted that adaptive management implies continuous refinement of the tools, models and collective understanding of the problem; conducting carefully designed monitoring of the results of each management action; and adjusting the overall solution set in response to continuous learning.

5.5.4 Optimization models

Various optimization techniques, in general, and their application to various hydrosystems problems, in particular, have shown remarkable progress over the past three decades. The progress of the application of these techniques has gone alongside the revolution of computer programs and, as such, similar explanations can be given to the development of simulation computer programs and optimization techniques over the past three or more decades. A wide variety of applications has been made using optimization techniques. These applications include surface water management, groundwater management, conjunctive use management, water distribution system operation, floodplain management, water supply systems, water quality management, urban drainage, stormwater management, and reservoir operation and management (see reviews in Mays (1997, 2002, 2004a, b, c, 2005) and Mays and Tung (1992).

Chapter 6

Opportunities and challenges

6.1 REALIZATIONS

There are many realizations that could be listed below. The following is only a short list of major realizations for integrated urban water management in arid and semi-arid regions of the world:

- When we discuss integrated urban water management there is a large disparity between what exists in developed countries and what exists in developing countries. In many arid and semi-arid countries, the existence of urban sewerage systems is limited where there are either no sanitation systems, or where only household pits and septic systems are available. In many areas where there are sewerage systems, the waste is discharged with little or no treatment, into rivers, lakes, wadis, the sea, or on open land or underground.
- Many urban water systems, particularly in developing countries, have low and/or erratic pressures due to inadequate supplies and/or operation concerns that allow contamination from groundwater and/or from sewerage systems that make the water not drinkable.
- Poor water management hurts the poor most. The Dublin principles aim at wise management with focus on poverty.
- The lack of adequate wastewater treatment, particularly in developing countries, is a serious threat to human health.
- There has been little or no integrated management between urban water excess management and urban water supply management, particularly in developing regions.
- On a global scale, there is probably enough water to meet both present and projected needs; however, the uneven geographical and temporal distribution, coupled with an increasing population that is often concentrated in physically water-scarce arid and semi-arid regions such as MENA, result in severe problems for the future of urban water management.
- Global climate change has been given very little consideration in integrated urban water management, even in highly developed arid and semi-arid regions, such as in the south-western United States.

- The reality is that climate change impacts are already with us and are manifesting in increasing occurrences of and intensity of climate extremes such as droughts and floods, and climate variability. From the water supply perspective, there are both supply-side and demand-side options that could be considered for urban water supply.
- Desalination of brackish and seawater has been expanding rapidly in recent times, primarily to provide water for municipal and industrial uses in arid and semi-arid areas. This has prevailed mostly in oil-rich Middle Eastern countries which have fuel supplies for the required energy.
- Artificial recharge is becoming an integral part of integrated urban water management, particularly in arid and semi-arid regions for sustainable water supplies, as it has many advantages over conventional surface water storage, especially in arid and semi-arid regions.
- Water reclamation and reuse has become an attractive option for conserving and extending available water supplies in arid and semi-arid regions, as many urban areas in these regions approach and reach the limits of their available water supplies. Water reuse can also provide urban areas with an opportunity for pollution abatement when effluent discharges are replaced to sensitive surface water.
- In arid and semi-arid regions, large water-transfer projects have been used as one solution to meet urban water demand because of the regional differences to water availability. However, projects that transfer water that is a non-renewable resource must be considered carefully, as these transfers cannot last forever.
- Rainfall harvesting can be used as a supplementary or even primary water source at the household or small community level, and rainwater harvesting can provide a significant amount of freshwater in locations such as the Arabian Gulf States.
- Water resources sustainability concepts must be incorporated into integrated urban water management.

6.2 GAMBLING WITH WATER IN THE DESERT

Urban water management has continuously been a challenge for agencies throughout the south-western United States; however, none has been as controversial as water management in Arizona. Each urban water supply system possesses different problems, but the one commonality is that they are all headed toward unsustainability with the rapid urban population growth; in other words, gambling with water in the desert. Most urban water supplies for this region come from various groundwater aquifers, surface reservoirs on the Salt River and the Central Arizona Project (CAP) water from the Colorado River (see Figures 3.13 and 3.14). The state water policies and laws coupled with federal policies related to interstate boundaries, Native American Indian issues and international boundaries make urban water management a continual challenge.

The concept of 'safe yield' or a balance between groundwater withdrawals and replenishment over time is the first part of the gamble. The concept of safe yield as defined in the Arizona Groundwater Management Act is not the same as the concept of sustainability. A second part of the gamble is that new urban development requires a demonstration that there is an 'assured water supply'. This concept supposedly prohibits growth that interferes with the ability to provide water of a quantity and quality necessary to satisfy regional water demand for the next 100 years, while meeting safe yield. In most urban areas in the state, water demand has outgrown the renewable groundwater supply, so that 'sustained growth', the third major gamble, is only possible through the CAP water from the Colorado River. The fourth gamble is the fact that Arizona has the lowest priority use of the CAP water, making it the state most vulnerable to shortages. During extended droughts, it would have no share. The fifth part of the gamble is that global climate change has not been given any consideration in the development of these other gambles. There is basically no consideration of water resources sustainability for the under-controlled urban growth, improperly referred to as sustained growth.

The major message is that even in very developed arid and semi-arid regions of the world, like Arizona, gambling with water for the sake of making money from growth, with little regard to sustainable development, exists on a large scale. Growth control may be one of the major answers, or in some cases the only answer, not only in the developing regions of the world, but also in the highly developed regions of the world. There are many other arid and semi-arid regions in the world that are gambling with water in the desert.

The overall goal of integrated urban water resources management should be water resources sustainability which has been defined as follows:

> Water resources sustainability is the ability to use water in sufficient quantities and quality from the local to the global scale to meet the needs of humans and ecosystems for the present and the future to sustain life, and to protect humans from the damages brought about by natural and human-caused disasters that affect sustaining life.

References

Adams, R. 1981. *Heartland of Cities: Surveys of Ancient Settlements and Land Use on the Central Floodplain of the Euphrates*. University of Chicago Press, Chicago.

Agnew, C. and Anderson, E. 1992. *Water Resources in the Arid Realm*. Routledge, London.

Alegre, H. 2002. Performance Indicators as a Management Support Tool. L.W. Mays (ed.) *Urban Water Supply Handbook*. McGraw-Hill, New York, pp. 9.3–9.74.

Alegre, H., Hirner, W., Baptista, J.M. and Parena, R. 2000. *Performance Indicators for Water Supply Services*. Manual of Best Practice Series. IWA Publishing, London.

Alley, W.M., Reilly, T.E. and Franke, O.L. 1999. *Sustainability of Ground-water Resources*. US Geological Survey Circular 1186. US Geological Survey, Denver, CO.

Arizona Department of Water Resources (ADWR). 1998. *Overview of Arizona's Groundwater Management Code*. www.adwr.state.az.us/Azwaterinfo/.

Arizona State University, University of Arizona, and University of Colorado. 1998. *Soil Treatability Pilot Studies to Design and Model Soil Aquifer Treatment Systems*. AWWA Research Foundation and American Water Works Association, Denver, Co.

Asano, T. (ed.) 1985. *Artificial Recharge of Groundwater*. Butterworth Publishers, Boston, MA.

Asano, T. 1999. Groundwater Recharge with Reclaimed Municipal Wastewater – Regulatory Perspectives. *Proceedings of the International Symposium on Efficient Water Use in Urban Areas – Innovative Ways of Finding Water for Cities*. UNEP IETC (International Environmental Technology Centre) Report 9, Osaka/Shiga, Japan, pp. 173–181.

Beaumont, P. 1971. Qanat Systems in Iran, *Bull. Intl. Assoc. Sci. Hydrology*, V. 16, pp. 39–50.

Bouwer, H. 1978. *Groundwater Hydrology*. McGraw-Hill, New York.

Broshi, M. 1980. The Population of Western Palestine in the Roman-Byzantine period, *Bulletin of the American Schools of Oriental Research*, 236, pp. 1–10.

Bruins, H.J. 2002. Israel: Urban Water Infrastructure in the Desert. L.W. Mays (ed.) *Urban Water Supply Handbook*. McGraw-Hill, New York, pp. 17.1–17.22.

Bull, W.B. 1988. Floods, Degradation, and Aggradation, V.R. Baker, R.C. Kochel, P.C. Patton (eds.) *Flood Morphology*, John Wiley and Sons, Inc., New York.

Cabrera, E. and Lund, J.R. 2002. Regional Water Management: A Long View. E. Cabrera, R. Cobacho and J.R. Lund (eds.) *Regional Water System Management: Water Conservation, Water Supply, and System Integration*. A.A. Balkema Publishers, Lisse, pp. 343–346.

Chepil, W.S. and Woodruff, N.P. 1954. Estimates of Wind Erodibility of Field Surfaces, *J. Soil Water Cons*. Vol. 9, No. 6, pp. 257–265.

Chow, V.T., Maidment, D.R. and Mays, L.W. 1988. *Applied Hydrology*. McGraw-Hill, New York.

Crook, J. 1998. *Water Reclamation and Reuse Criteria*. U.S. Environmental Protection Agency.

Cullinane, M.J. Jr. 1989. Methodologies for the Evaluation of Water Distribution System Reliability/Availability, Ph.D. Dissertation, University of Texas at Austin.

Dick-Peddie, W.A. 1991. Semiarid and Arid Lands: A Worldwide Scope, Semiarid Lands and Deserts. J. Skujins (ed.) *Soil Resource and Reclamation*. Marcel Dekker, Inc., New York.

Dregne, H.E. 1976. *Soils of Arid Regions*. Elsevier, Oxford.

Eheart, J.W. and Lund, J.R. 1996. Water-use Management: Permit and Water Transfer Systems. L.W. Mays (ed.) *Water Resources Handbook*. McGraw-Hill, New York, pp. 32.1–32.34.

Elgabaly, M.M. 1980. Problems of Soils and Salinity. A. K. Biswas et al. (eds.) *Water Management for Arid Lands in Developing Countries*. Pergamon Press, Oxford.

Evans, S.A. 1921–1935. *The Palace of Minos at Knossos: A Comparative Account of the Successive Stages of the Early Cretan Civilization as Illustrated by the Discoveries, Vols. I–IV*, Macmillan, London, vol. I., (reprinted by Biblio and Tannen, New York, 1964).

Feniera, V. 1990. Temporal Characteristics of Arid Land Rainfall Events. R. H. French (ed.) *Hydraulics/Hydrology of Arid Lands*. American Society of Civil Engineers, New York.

Figueres, C.M. 2005. Urban Water Management in the Middle East and Central Asia: Problem Assessment, *12th World Congress*, International Water Resources Association.

Friends of the Lower Olentangy Watershed (FLOW). 2003. *The Lower Olentangy Watershed Action Plan in 2003 – Strategies for Protecting and Improving Water Quality and Recreational Use of the Olentangy River and Tributary Streams in Delaware and Franklin Counties. Columbus, OH*. www.olentangywatershed.org/LowerOlentangyActionPlan04.pdf.

Fox, P. 1999. Advantages of Aquifer Recharge for a Sustainable Water Supply. *Proceedings of the International Symposium on Efficient Water Use in Urban Areas – Innovative Ways of Finding Water for Cities*. UNEP IETC (International Environmental Technology Centre) Report 9, Osaka/Shiga, Japan, pp. 163–172.

Frederiksen, H.D. 1997. Institutional Principles for Sound Management of Water and Related Environmental Resources. A.K. Biswas (ed.) *Water Resources – Environmental Planning, Management, and Development*. McGraw-Hill, New York, pp. 529–577.

Fuller, W.H. 1974. Desert Soils. G. W. Brown (ed.) *Desert Biology*, Vol. 2. Academic Press, London.

Gleick, P.H., P. Loh, S. Gomez and J. Morrison, 1995. *California Water 2020: A Sustainable Vision*. Pacific Institute for Studies in Development, Environment and Security, Oakland, California.

Goodrich, D.C., Woolhiser, D.A. and Unkrich, C.L. 1990. Rainfall-sampling Impacts on Runoff. R.H. French (ed.) *Hydraulics/Hydrology of Arid Lands*. American Society of Civil Engineers, New York.

Goudie, A. (ed.) 1985. *Encyclopaedic Dictionary of Physical Geography*. Blackwell, Oxford.

Grigg, N.S. 1986. *Urban Water Infrastructure – Planning, Management, and Operations*. Wiley-Interscience, New York.

Grigg, N.S. 1996. *Water Resources Management*. McGraw-Hill, New York.

Guo, J.C.Y. 2001. Design of Infiltration Basins for Stormwater, L.W. Mays (ed.) *Stormwater Collection Systems Design Handbook*. McGraw-Hill, New York.

Harper, J.K. and Griffin, R.C. 1988. The Structure and Role of River Authorities in Texas, *Water Resources Bulletin*, American Water Resources Association, Issue No. 6, Dec., Bethesda, Maryland.

Hassan, F.A. 1998. Climatic Change, Nile Floods, and Civilization, *Nature and Resources*, Vol. 32, No. 2, pp. 34–40.

Heathcote, R.L. 1983. *The Arid Lands: Their Use and Abuse*. Longman, London.

Hermann, T. and Schmida, U. 1999. Rainwater Utilization in Germany: Efficiency, Dimensioning, Hydraulic and Environmental Aspects, *Urban Water*, Vol. 1, No. 4, pp. 307–316.

Hillel, D. 1994. *Rivers of Eden: The Struggle for Water and the Quest for Peace in the Middle East*. Oxford University Press, Oxford.

Hills, E.S. 1966. *Arid Lands*. Methuen, London.

Howard, C.D.D. 2002. Sustainable Development-Risk and Uncertainty, *Journal of Water Resources Planning and Management*, Vol. 27, No. 5, pp. 309–311.

Intergovernmental Panel on Climate Change (IPCC). 2001. *Climate Change 2001: Impacts, Adaptation, and Vulnerability*. www.ipcc.ch.

Kahan, J.P., Wu, M., Hajiamiri, S. and Knopman, D. 2006. From Flood Control to Integrated Water Resource Management: Lessons for the Gulf Coast from Flooding in Other Places in the Last Sixty Years, Gulf States Policy Institute, Rand Corporation, Santa Monica, CA.

Khan, A.H. 1986. *Desalination Processes and Multistage Flash Distillation Practice*. Elsevier, New York.

Lane, L.J. 1982. Distributed Model for Small Semiarid Watersheds, *Journal of the Hydraulics Division*, ASCE, Vol. 108, No. HYlO, pp. 1114–1131.

Lane, L.J. 1990. Transmission Losses, Flood Peaks, and Groundwater Recharge. R.H. French (ed.) *Hydraulics/Hydrology of Arid Lands*. American Society of Civil Engineers, New York.

Lane, L.J., Diskin, M.H., Wallace, D.E. and Dixon, R.M. 1978. Partial Area Response on Small Semiarid Watersheds, *Water Resources Bulletin*, AWRA, Vol. 14, No. 5, pp. 1143–1158.

Laureano, P. 2005. *The Water Atlas: Traditional Knowledge to Combat Desertification*. Laia Libros, Barcelona, Spain.

Leeds, R., Brown, L.C. and Watermeier, N.L. 1993. *Nonpoint Source Pollution: Water Primer. AEX-465-93. Ohio State University Extension Fact Sheet*. Food, Agricultural and Biological Engineering, The State University of Ohio, Columbus, OH. http://ohioline.osu.edu/aex-fact/0465.html.

Logan, R.F. 1968. Causes, Climates, and Distribution of Deserts. G.W. Brown (ed.) *Desert Biology*. Academic Press, London, pp. 21–50.

Maia, R. 2006. Evaluation of Alternative Water Management Scenarios: Case Study of Ribeiras do Algarve, Portugal. P. Koundouri et al. (eds.) *Water Management in Arid and Semi-arid Regions*. Edward Elgar Publishers, Cheltenham, U.K., pp. 41–104.

Marsalek et al. 2006. *Urban Water Cycle: Processes and Interactions*. IHP-VI, Technical Publications in Hydrology, No. 78. UNESCO, Paris.

Marshall, J. 2005. Megacity, Mega Mess ..., *Nature*, 437, pp. 312–314, September 15.

Mays, L.W. (ed.) 1996. *Handbook of Water Resources*. McGraw Hill, New York.

Mays, L.W. 1997. *Optimal Control for Hydrosystems*. Marcel-Dekker, Inc., New York.

Mays, L.W. (ed.) 2000a. *Hydraulic Design Handbook*. McGraw-Hill, New York.

Mays, L.W. (ed.) 2000b. *Water Distribution Systems Handbook*. McGraw-Hill, New York.

Mays, L.W. (ed.) 2001. *Stormwater Collection Systems Design Handbook*. McGraw-Hill, New York.

Mays, L.W. (ed.) 2002. *Urban Water Supply Handbook*. McGraw-Hill, New York.

Mays, L.W. (ed.) 2004a. *Urban Water Supply Management Tools*. McGraw-Hill, Inc., New York.

Mays, L.W. (ed.) 2004b. *Urban Stormwater Management Tools*. McGraw-Hill, Inc., New York.

Mays, L.W. (ed.) 2004c. *Water Supply Systems Security*. McGraw-Hill, Inc., New York.

Mays, L.W. (ed.) 2005. *Water Resources Systems Management Tools*. McGraw-Hill, New York.

Mays, L.W. 2006. A Brief History of Ancient Water Distribution, *Proceedings, 1st IWA International Symposium on Water and Wastewater Technologies in Ancient Civilizations*, National Agricultural Research Foundation, Iraklio, Greece, pp. 27–36.

Mays, L.W. (ed.) 2007. *Water Resources Sustainability*. McGraw-Hill, New York.

Mays, L.W., Koutsoyiannis, D. and Angelakis, A.N. 2007. A Brief History of Urban Water Supply in Antiquity, *Water and Science Technology: Water Supply*, Vol. 7, No. 1, pp. 1–12, IWA Publishing.

Mays, L.W. and Tung, Y.K. 1992. *Hydrosystems Engineering and Management*. McGraw-Hill, New York.

McCann, B. 2006. Resource Risks of Climate Change, *Water 21*, IWA Publishing, London, October, pp. 16–18.

Meigs, P. 1953. World Distribution of Arid and Semi-arid Homoclimates, *Rev. Res. On Arid Zone Hydrol*. UNESCO, Paris, pp. 203–210.

Mitchell, B. (ed.) 1990. *Integrated Water Management: International Experiences and Perspectives*. Belhaven Press, London.

Morris, E.M. and Woolhiser, D.A. 1980. Unsteady One-dimensional Flow over a Plane: Partial Equilibrium and Recession Hydrographs, *Water Resources Research*, AGU, Vol. 16, No. 2, pp. 355–360.

Mortada, H. 2005. Confronting the Challenges of Urban Water Management in Arid Regions: Geographic, Technological, Sociocultural, and Psychological Issues, *Journal of Architectural and Planning Research*, Vol. 22, No. 1, Spring.

Nouh, M. and Al-Shamsy, K. 2001. Sustainable Solutions for Urban Drainage Problems in Arid and Semi-arid Regions. M. Nouh (ed.) *Urban Drainage in Arid and Semi-arid Climates, Vol. III*. UNESCO IHP-V, Technical Documents in Hydrology, No. 40, UNESCO, Paris, pp. 106–120.

Pantell, S.E. 1993. *Seawater Desalination in California*. California Coastal Commission, San Francisco, California. www.coastal.ca.gov/desalrpt/dchap1.html.

Pavelko, M.T., Wood, D.R. and Laczniak, R.J. 1999. Las Vegas, Nevada: Gambling with Water in the Desert. D. Galloway, D. Jones and S. Ingebritsen (eds.) *Land Subsidence in the United States*. US Geological Survey Circular 1182, pp. 49–64, , Denver, CO. http://water.usgs.gov/pubs/circ/circ1182.

Post, J. 2006. Wastewater Treatment and Reuse in the Eastern Mediterranean Region, *Water 21*, IWA Publishing, London, June, pp. 36–41.

Postel, S., *Last Oasis: Facing Water Security*, Norton, New York, 1997.

Reitsma, R.F. et al. 1996. Decision Support Systems (DSS) for Water Resources Management. L. W. Mays (ed.) *Water Resources Handbook*. McGraw-Hill, Inc., New York.

Renard, K.G. 1970. *The Hydrology of Semiarid Rangeland Watersheds, ARS 41–162*. US Department of Agriculture, Agricultural Research Service, Washington, D.C.

Roudi-Fahima, Creel, F.L. and De Souza, R.-M. 2002. *Finding the Balance: Population and Water Scarcity in the Middle East and North Africa*. Population Reference Bureau, Washington, D.C.

Schick, A.P. 1988. Hydrologic Aspects of Floods in Extreme Arid Environments. V.R. Baker, R.C. Kochel and P.C. Patton (eds.) *Flood Geomorphology*. John Wiley & Sons, New York.

Schmandt, J., Smerdon, E.T. and Clarkson, J. 1988. *State Water Policies, A Study of Six States*. Praeger, New York.

Sharon, D. 1972. The Spottiness of Rainfall in a Desert Area, *J. Hydrology*, Vol. 17, pp. 161–175.

Sharon, D. 1981. The Distribution in Space of Local Rainfall in the Namib Desert, *J. Climatology*, Vol. 1, pp. 69–75.

Shen, H.W. and Julien, P.Y. 1993. Erosion and Sediment Transport. D.R. Maidment (ed.) *Handbook of Hydrology*. McGraw-Hill, New York.

Shmida, A. 1985. Biogeography of the Desert Flora. M. Evenari, I. Noy-Meir and D.W. Goodall (eds.) *Ecosystems of the World, Vol. 12A, Hot Deserts and Arid Shrublands*. Elsevier Science Publishers, Amsterdam, pp. 23–77.

Sit, V. and Taylor, B. (eds.) 1998. Statistical Methods for Adaptive Management Studies, *Land Management Handbook. No. 42*. Res. Br., B.C. Min. For., Victoria, BC.

Skaggs, R.L., Vail, L.W. and Shankle, S. 2002. Adaptive Management for Water Supply Planning: Sustaining Mexico City's Water Supply. L.W. Mays (ed.) *Urban Water Supply Handbook*. McGraw-Hill, New York, pp. 15.3–15.24.

Slatyer, R.O. and Mabbutt, J.A. 1964. Hydrology of Arid and Semiarid Regions. Section 24 V.T. Chow (ed.) *Handbook of Applied Hydrology*. McGraw-Hill Book Company, New York.

Stedman, L., 2006. Assisting Africa, *Water 21*, IWA Publishing, London, August, pp. 34–36.

Strouhal, E. 1992. *Life in Ancient Egypt*. Cambridge University Press.

Tchobanoglous, G. 1996. Wastewater Treatment. Chapter 20, L.W. Mays (ed.) *Water Resources Handbook*. McGraw-Hill, New York.

Thomas, D.S. (ed.) 1989. *Arid Zone Geomorphology*. Belhaven Press, London.

Thompson, R.D. 1975. *The Climatology of the Arid World*, Paper No. 35, Department of Geography, Reading University.

Todd, D.K. and Mays, L.W. 2005. *Groundwater Hydrology*, 3rd edition. John Wiley and Sons, New York.

Todini, E., Schumann, A. and Assimacopoulos, D. 2006. The WaterStrategyMan Decision Support System. P. Koundouri et al. (eds.) *Water Management in Arid and Semi-arid Regions*. Edward Elgar Publishers, Cheltenham, UK, pp. 13–40.

Tung, Y.K. 2007. Uncertainties and Risks in Water Resources Projects. *Water Resources Sustainability*. McGraw-Hill, New York.

Tung, Y.K. and Yen, B.C. 2005. *Hydrosystems Engineering Uncertainty Analysis*. McGraw-Hill, New York.

Tung, Y.K., Yen, B.C. and Melching, C.S. 2006. *Hydrosystems Engineering Reliability Assessment and Risk Analysis*. McGraw-Hill, New York.

Urbonas, B.R. and Roesner, L.A. 1993. Hydrologic Design for Urban Drainage and Flood Control. Chapter 28, D.R. Maidment (ed.) *Handbook of Hydrology*, McGraw-Hill, Inc., New York.

US Environmental Protection Agency (EPA) and US Agency for International Development (AID). 1992. *Guidelines for Water Reuse*, EPA/625/R-92/004. Washington, D.C., September.

US Environmental Protection Agency (EPA) and US Agency for International Development (AID). 2004. *Guidelines for Water Reuse*, EPA/625/R-04/108. Washington, D.C.

Viollet, P.-L. 2006, Water Management in the Early Bronze Age Civilization, *Proceedings*, *La Ingenieria Y La Gestion Del Agua a Traves de Los Tiempos*, held at the Universidad de Alicante, Spain, with the Universidad Politechnica de Valencia, Spain.

Von Sperling, M. 1996. Comparison Among the most Frequently Used Systems for Wastewater Treatment in Developing Countries, *Water Science and Technology*, Vol. 33, No. 3, pp. 59–72.

Walters, C.J. and Holling, C.S. 1990. Large-scale Management Experiments and Learning by Doing, *Ecology*, Vol. 71, No. 6, pp. 2060–2068.

Waterstone, M., 1992, Of Dogs and Tails: Water Policy and Social Policy in Arizona, *Water Resources Bulletin*, American Water Resources Association, Vol. 28, No. 3, pp. 479–486.

Wehmhoefer, R.A. 1989. Chapter 2 in Z.A. Smith (ed.) *Water and the Future of the Southwest.*, University of New Mexico Press, Albuquerque, New Mexico.

Woolhiser, D.A. and Liggett, J.A. 1967. Unsteady, One-dimensional flow over a Plane – the Rising Hydrograph, *Water Resources Research*, AGU, Vol. 3, No. 3, pp. 753–771.

Woolhiser, D.A., Smith, R.E. and Goodrich, D.C. 1990. *KINEROS A Kinematic Runoff and Erosion Model: Documentation and Users Manual*. US Department of Agriculture, Agricultural Research Service, ARS-77.

World Water Assessment Program. 2006. *Water, a Shared Responsibility: The UN World Water Development Report 2*. UNESCO and Berghahn Books.

Zuhair, A., Nouh, M. and El Sayed, M. 1999. *Flood Harvesting in Selected Arab States*, Final Report No. M31/99, Institute of Water Resources, p. 217.

Case Study I

Water and wastewater management in Mexico City

Blanca Jimenez

Instituto de Ingeniería, Universidad Nacional Autónoma de México, Mexico

1 INTRODUCTION

Mexico City is the capital of Mexico, a country with around 103 million inhabitants (INEGI, 2005). Since Aztec times, Mexico City has been the most important city in the country. It comprises nearly 21 million people and produces 21% of GDP (Gross Domestic Product) due to its commercial, industrial and political activity. This intense activity combined with a huge population living within a closed basin high above sea level has created peculiar and complex water supply and disposal problems. This case study describes the management of water in the city. To explain the present situation, first historical aspects are presented followed by a description of water management activities, existing problems and present challenges. Because the disposal and future sources of water involve the management of water in the Tula Valley located to the north to Mexico City, its situation is also described. Finally, ways to improve Mexico City's water management are proposed that may be applicable to other megacities.

2 MEXICO CITY

2.1 Description

Mexico City lies in the Mexico Valley which is an endorreic (closed) basin of 9,600 km^2, located 2,240 metres above sea level, between 98°31′58″ and 99°30′52″ west longitude and 19°01′18″ and 20°09′12″ north latitude (Guillermo, 2000). The mean annual temperature is 15°C. Annual pluvial precipitation is 700 mm, varying from 600 mm in the north to 1,500 mm in the south. Pluvial precipitation normally occurs from May to October and is characterized by intense showers lasting short periods (Guillermo, 2000). One single storm, for instance, may produce 10–15% of the mean annual pluvial precipitation.

Mexico City was founded by the Aztecs in 1325 AD, who named it Tenochtitlan. When the Spanish arrived, in 1519 AD, Tenochtitlan was the most important city in Mexico, measuring 15 km^2 and having 200,000 inhabitants. The city was an island surrounded almost entirely by water, with only four streets providing communication with the land (Figure 1a). These streets were also dikes; the biggest called Nezahualcóyotl, was 12.6 km long and 6.7 m wide. Mexico City has now expanded to 8,084 km^2. Initially, there were five lakes in the valley, some of them communicating

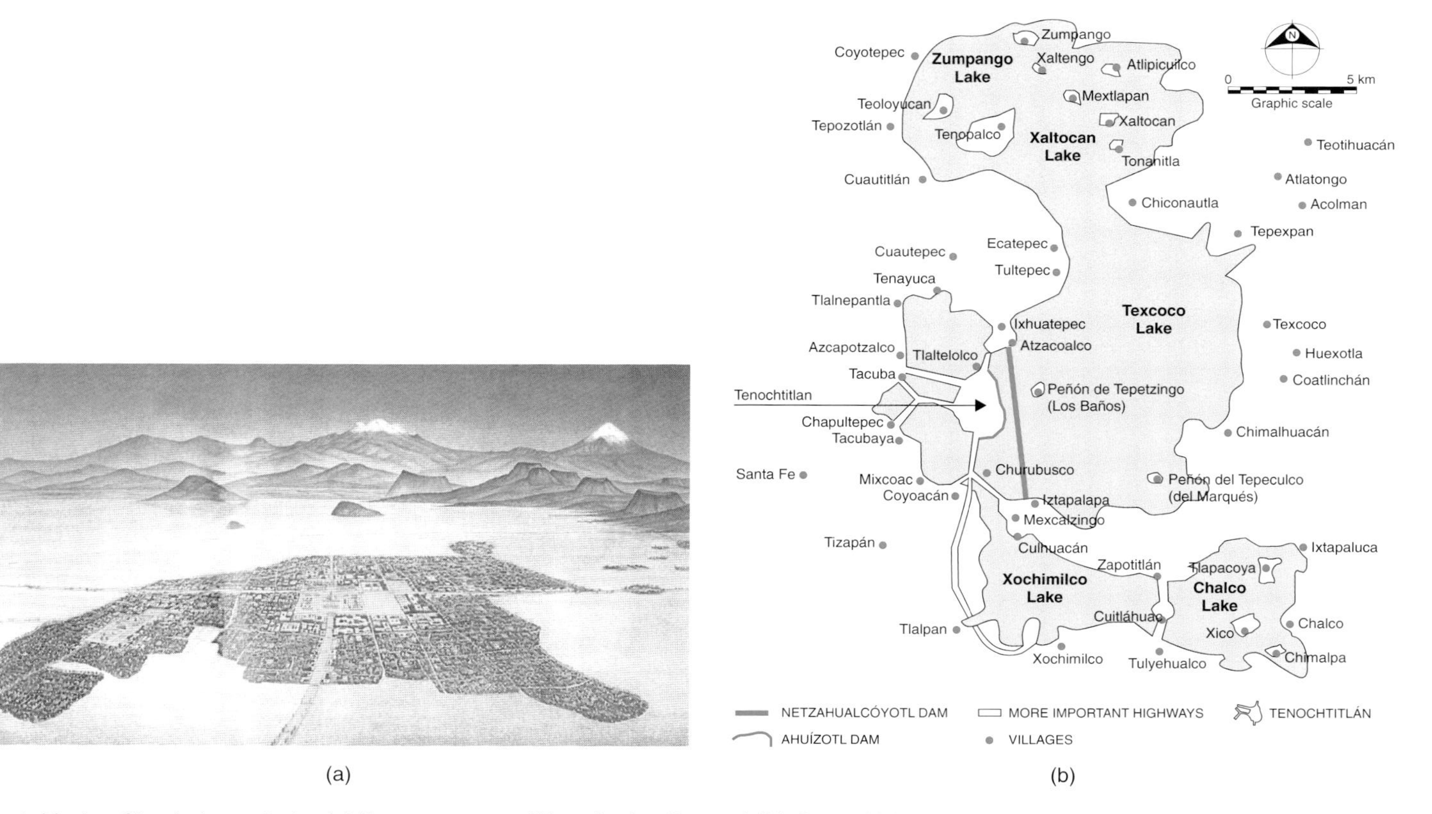

Figure 1 Mexico City during antiquity **(a)** Representation of Tenochtitlan City; and **(b)** Original lakes in the Mexico Valley (See also colour plate 11)

Source: Santoyo et al., 2005

with each other: Zumpango, Xaltocan, Texcoco, Xochimilco and Chalco (Figure 1b). The biggest lake was also the lowest in altitude; it did not directly receive water from springs and was fed with water from the other lakes, and therefore its salinity was higher. During heavy rains, the five lakes formed a single lake. Due to urban expansion and the artificial desiccation of the valley, today only small parts of the Texcoco and Zumpango lakes remain. Both are fed with pluvial excess water plus treated wastewater in the case of the Lake Texcoco.

In the Mexico Valley there are very few perennial rivers as most carry water only during the rainy season. Rivers were born in the south and west of the city that were used as sewers during the Spanish conquest and were finally converted to covered sewage canals after the 1950s.

Initially Mexico City was only located in the Federal District, but it now includes 37 adjoining municipalities from the State of Mexico. Currently, there are more people living in the State of Mexico (around 60% of the population) than in the Federal District.

2.2 Water system

2.2.1 History

During the Aztec period, excess water rather than a lack of water was a problem. Tenochtitlan City was designed to manage floods by controlling the water level in lakes and canals (used also as a means of communication) through a complex set of sluices. When the Spanish colonized the city, they did not understand the hydraulic system, which led to its destruction. As a result, the city frequently suffered both floods and lack of water. As the city grew, water began to be imported from other sites in the valley, and in 1847, when the population grew to around 0.5 million people, local groundwater began to be exploited (Garza and Chiapetto, 2000) from wells at a 105 m depth (Santoyo et al., 2005). By 1857, there were 168 wells that produced artesian water. Later, in 1942, when the city had grown to almost 2 million, water had to be imported from the Lerma River, in a basin located in the State of Mexico, and, in 1951, from the groundwater of this same basin. Finally, in 1975, when the population had reached 7 million, surface water was imported from the Cutzamala region, 130 km away and 1,100 m below Mexico City's level.

2.2.2 Water supply

At the present time, Mexico City, with 21 million inhabitants, uses 85.7 m^3/s of water (Figure 2), 48% of which is supplied through the network system, 19% is directly pumped from a local aquifer by farmers and industries, and the rest, 9%, corresponds to treated wastewater that is being reused (Capella, 2006 and Jiménez et al., 2004a). First use water (78 m^3/s) comes from the following sources:

- 57 m^3/s from 1965 wells that are extracting from the local aquifer located mainly south and west of the city
- 1 m^3/s from local rivers located in the southern part of the city
- 5 m^3/s from an aquifer located in the Lerma region, located 100 km away and 300 m above the Mexico City level
- 15 m^3/s from the Cutzamala river system, located 130 km away and 1,100 m below the Mexico City level.

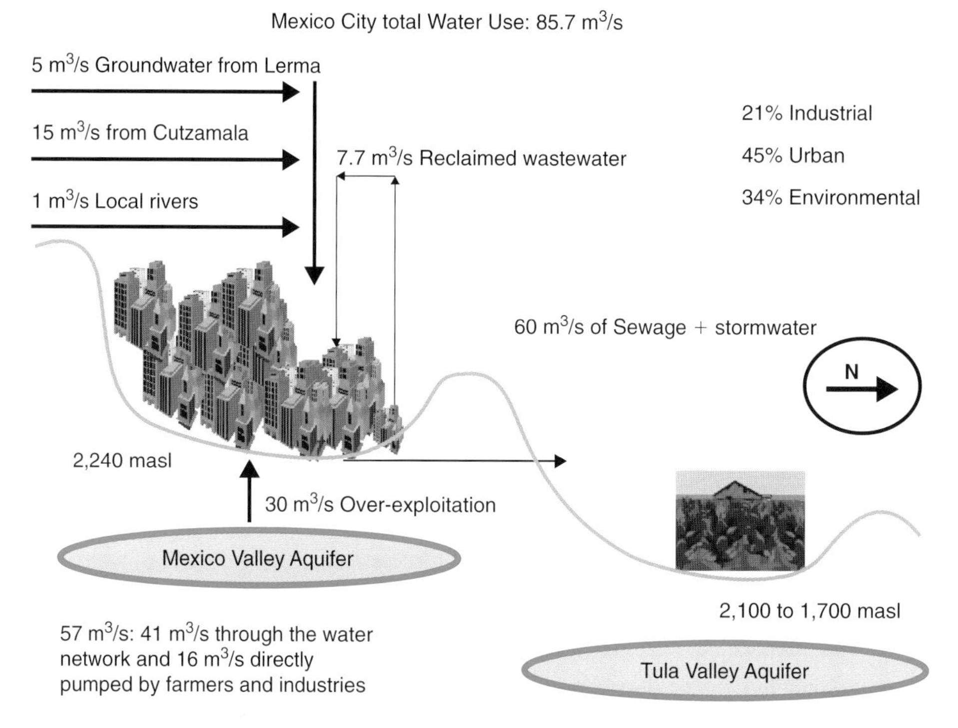

Figure 2 Water sources of Mexico City (See also colour plate 12)

Water service coverage is 89% in the Federal District and slightly lower in the municipalities of the State of Mexico. There is no data about the population receiving a regular service in the State of Mexico, but in the Federal District alone there are 1.15 million people receiving water intermittently through water tanks[1] (SACM, 2006).

2.2.3 Water uses

Water is used for municipal purposes (74%), followed by fresh water irrigation (16%), self-supplied industries (2%) and for non-drinking-water reuses (1%). Besides human consumption, municipal uses include commercial and low water consuming industries. Agriculture takes place in 40,000 ha of the valley. Reuse is performed mainly for municipal purposes (lake refilling, park and garden irrigation, car washing, environmental restoration and fountains), followed by industrial uses.

2.2.4 Actual use of water

It is often thought that because 62 m^3/s of water is distributed to the population through the water network, water use *per capita* per day is 255 L. However, 40% of the water leaks from the network, so people only receive 153 L/*capita*.d, a value that

[1] Twice per week and at no cost.

Table 1 Water use by social class in Mexico City (DGCOH, 1998)

Social class	Water supply, L/capita.d	% of the population	Total volume of water demanded
Low	128	76.5	9.0
Medium	169	18.0	15.7
Medium High	399	3.6	1.3
High	567	1.9	1.0
		TOTAL	$27\,m^3/s$

falls perfectly within the range of 150–170 L/*capita*.d recommended by WHO (1995). In fact, the actual use of water by the population varies according to social class, as presented in Table 1, for the Federal District. The upper class, i.e. 5% of the population, uses more than four times the amount used by the lower class. This is important because instead of spending money on water saving media campaigns, it would be wiser for the government to charge people using excessive amounts of water with a higher tariff and use the revenue to repair leaks.

2.2.5 Management

Water is managed in the Federal District by the Water System, both publicly and privately, while the 37 municipalities of the State of Mexico, with a population 15% higher than that of the Federal District, manage water publicly and independently of each other. This represents a complex problem for integrated water management. In the Federal District, there are users in 2.1 million households, 1.8 million of which are registered and 1.3 million metered. Only around 60% of the users are charged based on consumption. The total commercial efficiency of the system (billing and charging) is 68%. There is no equivalent data for the municipalities of the State of Mexico.

2.2.6 Water quality

Water sources – There is little available information on the quality of water sources, and most of it relates to the local groundwater. Different studies reveal that groundwater is increasingly being polluted. In the western part of the valley, the TSS content has increased from 1,000 mg/L to 20,000 mg/L; the sodium content from 50–100 mg/L to 600–800 mg/L; the ammoniacal nitrogen content from 0–0.03 mg/L to 6–9 mg/L; and the iron content from < 0.1 mg/L to 3–6 mg/L (Bellia et al., 1992; DDF, 1985; Lesser et al., 1986; and Ezcurra et al., 2006). In very specific areas in the western part of the city, where uncontrolled dumping sites used to exist, over-exploitation is causing leachates to move into the aquifer. A high content of a wide variety of organic compounds has been reported and Fe and Mn content in the same south-western area have increased due to over-exploitation (Lesser et al., 1986; Saade Hazin, 1998). In the southern area, nitrates, ammoniacal nitrogen and fecal coliform content is increasing, in this case because there is no sewerage, instead houses discharge to septic tanks with no control and the effluent is sent directly to volcanic soil for its infiltration.

Table 2 Percentage of samples fulfilling the drinking-water limit of free residual chlorine and the bacteriological content in the water supplied in Mexico City (GDF, 1999)

Parameter	*1988*	*1989*	*1990*	*1991*	*1992*	*1993*	*1994*	*1995*	*1996*	*1997*	*1998*
Free residual chlorine	85	88	92	94	94	94	93	94	94	91	87
Bacteriology	70	82	84	88	91	93	93	92	92	91	83

Groundwater pollution is caused by over-exploitation, domestic and industrial discharges to the aquifer, uncontrolled non-point source pollution as well as seepage from septic tanks and sewerage. Wastewater arrives in the aquifer from two main sources: the deterioration of the infrastructure due to sharing forces provoked by soil subsidence; and the formation of soil fissures created by subsoil desiccation due to over-exploitation. Additionally, in the southern part of Mexico City, the lack of sewerage and the presence of at least half a million septic tanks discharging directly into a volcanic soil with high transmissibility is polluting the aquifer. According to Capella (2006), this phenomenon alone is recharging the aquifer with 1 m^3/s of sewage.

Drinking water – In Mexico City tap water is not safe to drink, as is the case in the rest of the country. To follow up, drinking-water quality, fecal coliform and free residual chlorine are the only parameters that are systematically measured. Official data about drinking-water quality is scarce (Table 2) and non-public. It was public only for a short period from 1988 to 1998, when the quality began to deteriorate (Table 2). From the highest quality reached in 1992–1996 when the number of samples fulfilling the drinking norm of free residual chlorine content and fecal coliform content was the highest (92–94%), it diminished to 87% and 83%, respectively in 1998.

Detailed data about drinking-water quality produced from the local aquifer exists, but unfortunately only in some isolated academic studies. Mazari-Hiriart et al., 2002, for instance, found nitrates, chloroform, bromo dichloro methane and total organic carbon in chlorinated water with a higher concentration during the dry season. The total trihalomethane content, however, did not surpass the Mexican drinking-water norm (NOM-127-SSA1-1994, DOF, 2000) of 200 μg/L but exceeded the maximum allowable value of 80 μg/L established in the United States (EPA, 2004). Concerning microbiological quality, these same authors reported the presence of total coliforms, fecal coliforms, fecal Streptococcus, and other pathogenic bacteria before chlorination and also afterwards. They isolated 84 micro-organisms of nine genera associated with human fecal pollution. Some of these were *Helicobacter pylori* (associated with gastric ulcers and cancer) and coliphages MS-2 (Mazari-Hiriart et al., 1999 and 2001).

With respect to the quality of the drinking water produced from surface sources, (Lerma, Cutzamala and Mexico Valley rivers), there is almost no information, although water sources are evidently becoming polluted.

To potabilize water, chlorination is used for groundwater sources while alum coagulation, sedimentation and chlorination are applied to surface water sources. All these processes were selected based on characterizations performed several decades ago, and until now no review of their appropriateness for addressing water quality problems has been performed.

Additionally, it has been shown that water quality deteriorates during distribution (Jiménez et al., 2004b). The water network operates intermittently at low pressure due to a lack of water supply across the city. To have water all day, people must use individual storage tanks and as a result, tap water is of a lower quality. To have drinking water, a family of four earning four times the minimum wage spends 6–10% of its income on bottled water or potabilizing tap water at home (by boiling it, adding disinfectants or having individual disinfection systems using ozone, UV-light or Silver colloids[2]).

2.2.7 Water and health

The diarrhoeic diseases morbidity rate in Mexico City is 5,606 cases for 100,000 inhabitants while for the whole country it is 4,971 for 100,000 (SSA, 2005). Cifuentes et al., (2002) found that there was no relation between the fecal coliform content and the enteric disease rate because this indicator was not modelling the actual water health risks well. They also found that an intermittent supply of water and the need to store water at home was the cause of an increase in enteric diseases, especially among the poorest sector of the population and particularly among children under five years of age. Diarrhoeic diseases show a seasonal pattern. Those caused by *Escherichia coli* and *Shigella* have higher morbidity rates in the rainy and hot season (April–September) while those caused by rotaviruses are higher in the dry and cold season (October–February) according to López-Vidal et al., 1990; LeBaron et al., 1990; and Guerrero et al., 1994.

2.2.8 Over-exploitation of local groundwater and water transfer from other basins impact

Over-exploitation – While natural water availability in the Mexico Valley aquifer is 24 m^3/s, actual exploitation is 52 m^3/s resulting in an over-exploitation rate of 117%. At first, over-exploitation led to water being imported from other basins, but this was eventually no longer feasible due to the cost and for political reasons. Since 1964, new wells have been added each year to respond to an increasing water demand from a population that is partly – and unwisely – growing south and east of the city, above the natural recharging area. As a result of over-exploitation, Mexico City is also suffering soil subsidence with sinking rates varying according to the area (Figure 3). The highest sinking rate registered was 40 cm/year, and in 1954 it was decided to close several wells located in the centre of the city and open new ones in the south, where the soil is volcanic, and in the north (Santoyo et al., 2005 and Ezcurra et al., 2006). However, by that time the centre had already sunk 7 m (Figure 4). The redistribution of wells reduced the sinking rate in the centre, but the western part of the city is still sinking at a rate of 25 cm/yr.

Mexico City soil subsidence is creating major problems for the urban infrastructure, including:

- A loss of the sewerage/drainage capacity. The drainage capacity of the Gran Canal built to convey most of Mexico City's wastewater has been dramatically reduced. This canal was initially built to transport a maximum of 200 m^3/s, but its capacity has since been reduced to dozens of cubic metres. A pumping station was recently built at a cost of 30,000 US$ (US dollars) to raise 40 m^3/s of wastewater a height of 30 m year

[2] Individual disinfection systems double the price to disinfect water.

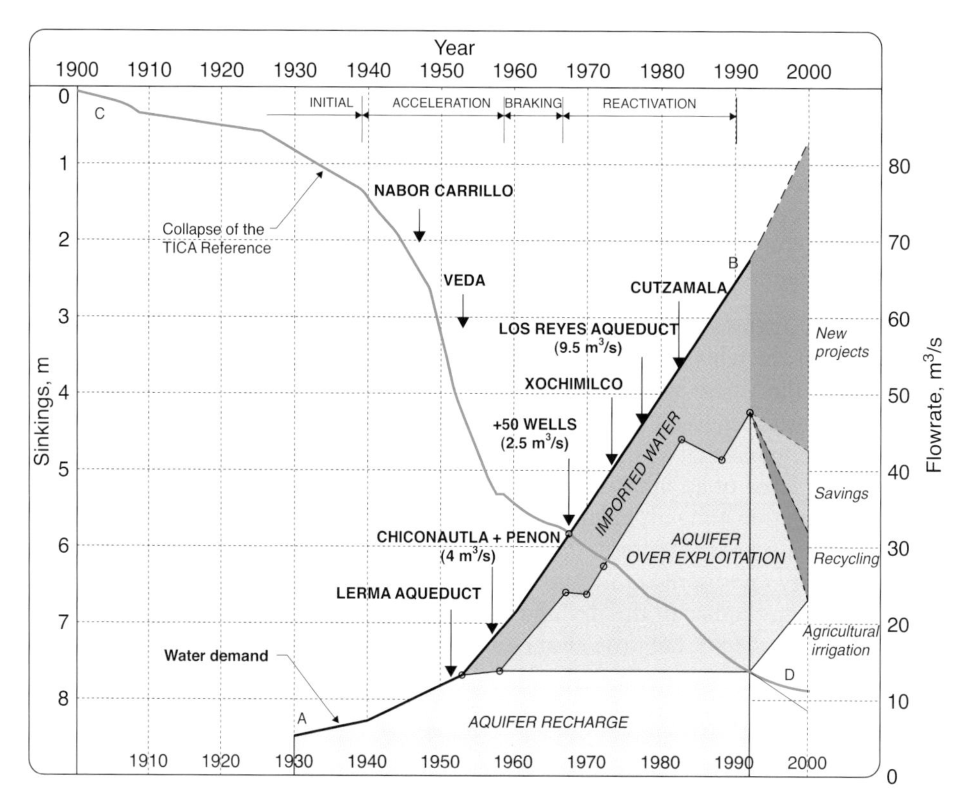

Figure 3 Sinking rates of Mexico City (See also colour plate 13)

Source: Santoyo et al., 2005

round (Domínguez, 2006). The Deep Drainage (western and central interceptors) built to convey 200 m^3/s of mainly pluvial water for half of the year, is transporting up to 300 m^3/s wastewater and pluvial water year round. This is possibly the most critical problem, because the Deep Drainage has not received maintenance for more than 10 years, and if it fails it will cause the centre of the city (an area of approximately 400 km^2) to be flooded with 1.2 m of wastewater affecting at least four million people (Domínguez et al., 2005). Due to a loss of drainage capacity, there are around 20–30 big floods (of more than 30 cm) per year causing non-quantified damage of varying magnitudes and nature (SACM, 2006) in the city (see Figure 5). The investment needed to recover wastewater drainage capacity in the Gran Canal alone is 305 million US$.

- Serious structural problems in buildings. The total number of damaged buildings and houses has not been documented, but in a small area of the historic downtown there are some 46 damaged structures (Santoyo et al., 2005). Due to its historic value, the Mexico City Cathedral is being repaired. It is estimated that in the last 50 years, differential sinking has led to an 87 cm difference between the apse and the western bell tower. The total investment required to redress the situation was 32.5 million US$ in 2000 (Santoyo and Ovando, 2002).
- Leaks in water and wastewater networks. Differential sinking leads to faults in water and sewerage pipelines producing leaks. Usually 37–40% of the water

Figure 4 Comparison of subsidence at a well casing **(a)** 1950 and **(b)** 2005

Source: Santoyo et al., 2005

conveyed is leaked, resulting in a water loss of 23 m^3/s. Only half of that volume would suffice to supply water to 3.2 million people at a rate of 300 L/*capita*.d. Sewerage leaks have not been evaluated.

- Due to over-exploitation, groundwater quality is deteriorating as explained above.
- Due to soil subsidence, the metro rails must be re-levelled each year, and in some parts, accumulated changes are compromising its operation.

Unfortunately, there is no hard data on the total cost of the effects of soil subsidence.

Water transfer from other basin – Importing water from the Lerma region, which used to be a lake, has caused fishing to disappear and people have had to live off the land. Furthermore, locals have gone from using surface water as a supply source to using water from wells. Ezcurra, et al., 2006, also points out that Chapala Lake, which was fed partly by the Lerma system, has experienced a reduction in level of 5 m. The transfer of water from the Cutzamala region reduced the amount of water available for power generation and caused the loss of a large irrigation area.

2.2.9 Future water demand

To continuously supply water to Mexico City's entire population there is a need for 1–2 m^3/s of water and to supply demand for an estimated population growth in the next five years, an additional 5 m^3/s are needed. Additionally, to prevent over-exploitation and for the injection of water to control soil subsidence, at least 15 m^3/s of water are required. It has been estimated that by the year 2010, 38 m^3/s of water will be needed. Part of this volume could certainly come from a leakage control

Figure 5 Freeway flooded in Mexico City in 2006

programme. In a five year programme, 10–15 m^3/s of water could be saved by investing 1.5 million US$ to sectorize and control pressure in the water network, plus 500 million US$ per year to repair and change deteriorated pipelines[3] (Capella, 2006). Considering the costs, the time needed to control leaks and recover water and the total volume required, other sources of water are needed, which could consist of the transfer of surplus water from other basins and/or the implementation of water reuse programmes.

2.2.10 Wastewater

Sewerage system – Because Mexico City is located in a closed basin (which used to be a lake, as explained) three artificial exits were built to drain wastewater and pluvial water. The work began in 1604, when the 11th Viceroy, Luis de Velazco, decided to desiccate the valley, drain the lakes and get rid of wastewater, and ended[4] in 1975. It was designed by Heinrich Martin, a German who Mexicanized his original name to 'Enrico Martinez'. The first important construction was the 'Tajo de Nochistongo' (Figure 6), which was hand dug by 29,650 Indians. At the entrance, the earth channel was 6.3 km long and 11 m deep; it was followed by an earth tunnel of 0.6 km, with 42 shafts, the deepest one measuring 22 m, and at the exit was 0.6 km long. When the tunnel collapsed, thousands of Indians were killed.

Later, the Viceroy commissioned Andrian Boot (from the Netherlands) and Enrico Martinez to present 'definitive' projects for resolving Mexico City's drainage problem. To decide whose design was the best, he ordered the closure of the Tajo de Nochistongo, producing in 1629, the worst flood (2 m in depth) ever to befall Mexico

[3] The total time needed to repair all the network is 50 years.
[4] Due to soil subsidence a new deep sewer needed to be built.

Figure 6 'Tajo de Nochistongo' during the Spanish period

City. The city remained flooded for five years. Around 20,000 people died, either due to the flood or as result of a pest epidemic. The issue of whether the capital of the country should be moved to the city of Puebla (Santoyo et al., 2005) was even raised. Enrico Martinez was put in jail but after being found not guilty of the flood, his project was retained. Subsequently, floods were controlled in a reasonable manner until the end of the nineteenth century with the construction of the Gran Canal in 1900. It went from San Lazaro (near the centre of Mexico City) to Zumpango (at the end of the valley) and had a proper slope to drain stormwater from the city. Later, in the 1960s, the Deep Drainage ('Drenaje Profundo') was built to drain only excess pluvial water from the city.

The drainage system now covers 94% of the Federal District and 85% of the municipalities of the State of Mexico (Merino, 2000). The drainage system conveys sewage and stormwater and is very complex. The sewerage system comprises 10,400 km of pipelines 0.3–0.6 m in diameter; 2,369 km of pipelines (76 km with a 3.05 m diameter); 96 pumping stations with a total capacity of 670 m^3/s; 91 underpasses for 14.3 m^3/s; 106 marginal collectors; 12 storm tanks with a total capacity of 130,000 m^3; several inverted siphons to overpass the Metro; three rivers; 29 dams; the Gran Canal (47 km long); and the Deep Drainage, 155 km long, with diameters of 3–6.5 m and a depth varying from 20 to 217 m.

Wastewater quality – Table 3 presents the composition of Mexico City's wastewater during the rainy and dry seasons. As shown, stormwater does not dilute the pollution. Actually, the TSS and helminth ova content are higher during the rainy season than during the dry one. Also, and contrary to what sanitary engineers hitherto believed, the metal content in the untreated wastewater is not very high, thus meeting thresholds on its use for irrigation without any treatment. This is due to the large amount of household wastewater produced that dilutes the relatively small volume of industrial discharges.

Table 3 Characterization of Mexico City's wastewater (Jiménez, 2005 and Jiménez et al., 1997)

Parameter (mg/L unless other unit is indicated)	*Dry season*			*Rainy season*		
	Mean	*Min*	*Max*	*Mean*	*Min*	*Max*
Chemical oxygen demand	527	245	1492	475	168	1581
Biochemical oxygen demand	240	20	330	180	40	420
Total Suspended Solids	295	60	1500	264	52	3383
Total Kjeldhal Nitrogen	26	18	47	17	2	61
Total Phosphorus	10	1	19	8.3	0.2	27
Helminth ova, eggs/L	14	6	23	27	7	93
Fecal Coliforms, MPN/100 mL	4.9×10^8	1.2×10^8	5.2×10^9	7.4×10^8	7.1×10^7	2.4×10^9

	Metals content			
	Mean	*Min*	*Max*	*Mexican standard*
Arsenic	0.004	0.001	0.006	0.2
Cadmium	0.005	0.004	0.006	0.05–0.4
Chromium	0.048	0.006	0.02	0.5–1.5
Copper	0.019	0.006	0.05	4–6
Lead	0.02	0.02	0.02	0.5–10
Nickel	0.16	0.11	0.18	2–4
Zinc	0.21	0.06	0.49	10–20

Wastewater treatment infrastructure – Mexico City produces, on a year round basis, 67.7 m^3/s of wastewater, a volume that includes pluvial excess water – which in fact is collected only six months per year – and sewage. There are several wastewater treatment plants in the City: 27 are operated by the Federal District Government, 44 by federal institutions (Ex Texcoco Lake Commission, Federal Electricity Commission and the army) or private owners and 20 by the municipalities of the State of Mexico (Merino, 2000). The public wastewater treatment plants' total capacity is 15 m^3/s, but only 7.7 m^3/s of the wastewater (11% of the total produced) is treated. All of the treated wastewater is reused.

Wastewater reuse began when the first wastewater treatment plant was installed in 1956 to irrigate green areas (Merino, 2000). Treated wastewater is now reused to fill recreational lakes and canals (54%) (Figure 7a), irrigate 6,500 ha of agricultural land and green areas (31%), for cooling in industry (8%), diverse purposes in commercial activities (5%) and to recharge the aquifer (2%) (DGCOH, 1998). The total amount of wastewater treated privately is not known; however, it is known that all of the wastewater is reused for lawn irrigation or cooling in industries. Considering that 100% of the treated wastewater is reused, or 12% of the wastewater produced, Mexico City is among the world's most intensive reuser of wastewater (Jiménez and Asano, in press)[5].

One of the biggest public reuse projects is the Ex-Texcoco Lake wastewater treatment plant. This plant, built at the beginning of the 1980s, has a 1 m^3/s capacity, but

[5] This even without considering that also 100% of the non-treated wastewater is also reused, as will be presented later in the text

Figure 7 **(a)** Chapultepec Recreational Lake; **(b)** Birds in the Texcoco Lake; and, **(c)** Dust storm in Mexico City before the Texcoco Lake was recovered (See also colour plate 14)

it only treats 0.6 m^3/s of wastewater due to civil construction problems. Originally, the intention was to exchange groundwater used for agriculture with reclaimed wastewater. The project consists of an activated sludge treatment plant followed by an artificially built lake of 1,380 ha to store and improve water quality. Treated wastewater is successfully used to refill the lake creating an environment where a wide variety of birds from Canada and the USA live during the winter (Figure 7b). Recovering part of the Texcoco Lake was very important to controlling the alkaline dust storms that the City (Figure 7c) frequently suffered and which were created by the wind carrying the fine dust that formed on the bottom of the ancient lake. Unfortunately, a high evaporation rate in the area and the solubilization of the salt contained in the soil considerably raised the effluent's salinity, making the water unsuitable for use in irrigation.

Wastewater disposal – Three tunnels were built to drain the pluvial water and the untreated wastewater out of the Mexico Valley to the Tula Valley: the Gran Canal in 1898, the Western Interceptor in 1896 and the Central Interceptor in 1975, these last two being part of the Deep Drainage (Figure 8). Although, the average annual volume of water disposed is approximately 60 m^3/s (80% sewage and 20% stormwater), wastewater flow actually varies from 52 to >300 m^3/s.

3 THE TULA VALLEY

3.1 Description

The Tula Valley is located 100 km north of Mexico City in the State of Hidalgo, between latitudes 19°54′ and 20°30′ north and longitudes 99°22′ and 98° 56′ west. The altitude varies from 2,100 m in the southern part to 1,700 m in the northern part. Soil types are: rendzic and melanic Leptosols; calcic and haplic Phaeozems; and eutric Vertisols (Siebe, 1998). Pluvial precipitation is 525 mm and occurs only five months of the year. In contrast, the annual evaporation rate is 1,750 mm. This low pluvial precipitation combined with a relatively high evaporation rate is why irrigation at high lamina rates is required in the valley to raise crops.

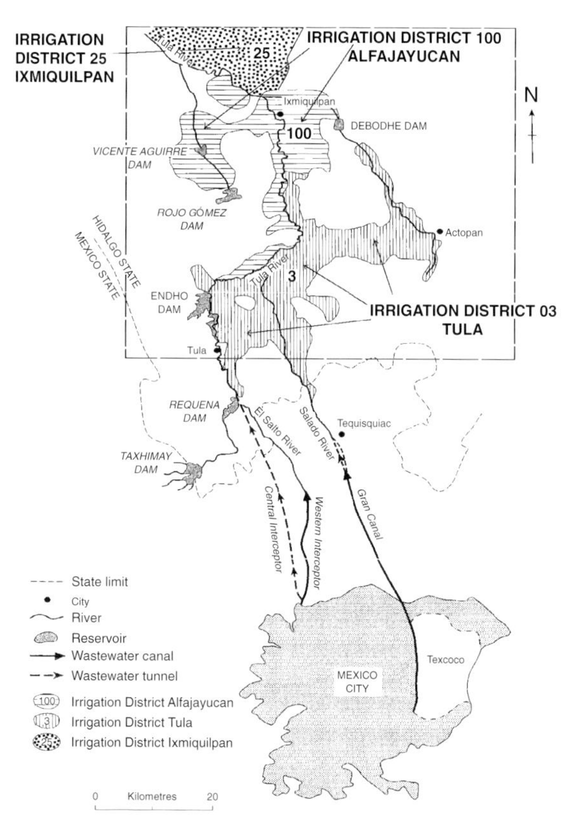

Figure 8 Mexico City's wastewater disposal drainage system and main components of the irrigation system in the Tula Valley

Initially, wastewater was disposed in the poorest and remotest area of the Tula Valley ('El Mezquital Valley'). Original vegetation was *Xerophila* scrubs, such as mezquite, sweet acacia, yucca and a wide variety of cactus (Siebe, 1998). The population disliked the decision, but in 1920, when an increase in agricultural production became evident, farmers requested that the government send more wastewater. Later they asked the President to grant them Mexico City's wastewater, which was done in 1955, when 26 m^3/s of wastewater were allocated to the 03 Tula District. A complex irrigation system was subsequently built (Figure 8). This now comprises nine dams (three for freshwater and six for wastewater); three rivers (Tula, Actopan and El Salado); and 858 km of unlined canals (Jiménez, 2005). One of the dams is the Endho Dam, which, with a capacity of 202,250 hm^3 (202.25 km^3), is probably one of the world's biggest wastewater dams.

Because Mexico City's growth produced increasing amounts of wastewater, the irrigated area gradually increased. In 1926 it was 14,000 ha, but by 1950 it had increased to 28,000 ha. In 1965 it measured 42,460 ha and in 1992 it had reached a maximum of 90,000 ha. Today, the irrigated area has decreased to 76,119 ha due to farmer migration to the United States. The irrigated area is divided into four neighbouring

Table 4 Yield increase resulting from irrigation using untreated wastewater (Jiménez, 1996)

Crop	*Yield, tons/ha*		*Increase, %*
	Wastewater	*'First use' water*	
Corn	5.0	2.0	150
Barley	4.0	2.0	100
Tomato	35.0	18.0	94
Forage oats	22.0	12.0	83
Alfalfa	120.0	70.0	71
Chilli	12.0	7.0	70
Wheat	3.0	1.8	67

irrigation districts: the 03-Tula, the 100-Alfajayucan, the 25-Ixmiquilpan and the 88-Chiconautla, this latter one in the State of Mexico (Jiménez, 2005). The region is globally and colloquially known as the Mezquital Valley, and has been considered by Mara and Cairncross (1989) as the largest irrigated area using wastewater in the world. Thanks to the wastewater, agriculture is the main economic activity. Corn and alfalfa that are used as fodder are the main crops (60–80% of the area) followed by oats, barley, wheat and some vegetables (chilli, Italian zucchini and beetroot). An important part of the produce is sold in Mexico City.

3.2 Wastewater use effects on agriculture and health

The use of wastewater improved the local economy, agriculture being the main source of income. This was not only due to the use of the wastewater to irrigate, but also to its nutrient and organic matter content. Table 4 shows a comparison of the yield obtained in two areas of the same soil, one irrigated with fresh water and the other with wastewater (Figure 9). Wastewater's economical value is recognized as increasing land prices (1 ha of land where wastewater is available can be rented at 455 US$/year compared to 183 US$/year for land that is rain fed, (Jiménez, 2005)).

In spite of improved agricultural production and improved welfare, serious health problems are also created. Cifuentes et al. (1992) showed that diarrhoeal diseases caused by helminths (worms) are increased by 16 times in children under 14 years (Table 5) but no increase was observed for protozoa (like *Entamoeba* and *Shigella*) which have another infection pathway different to the use of wastewater in irrigation. Results presented in Table 5 have actually been used by the WHO (1989 and 2006) to establish criteria for limiting the pathogenic content in the reuse of wastewater for agricultural irrigation.

Diseases caused by helminths are common only in developing countries. Globally there are five million people suffering from helminthiasis (UN, 2003). The use of wastewater for irrigation or fish production is the main pathway of infection (WHO, 2006). There are several kinds of helminthiasis, ascariasis being the most common and endemic in Africa, Latin America and the Far East. Helminths are transmitted through infective eggs (Figure 10) contained in crops irrigated with wastewater.

Figure 9 El Mezquital area **(a)** with and **(b)** without wastewater for irrigation (See also colour plate 15)

Table 5 Comparison of the frequency of water-borne illnesses in the Mezquital Valley and an area using clean water, with information from: Cifuentes et al. (1992)

Species	*Affected pop. by age*	*Death rate*		
		Area irrigated with wastewater (A)	*Area irrigated with clean water (B)*	*Ratio (A/B)*
Ascaris	0 to 4	15.3	2.7	5.7
lumbricoides	**5 to 14**	**16.1**	**1.0**	**16.0**
(Helminth)	>15	5.3	0.5	11.0
Giardia	0 to 4	13.6	13.5	1.0
lamblia	5 to 14	9.6	9.2	1.0
(Protozoon)	>15	2.3	2.5	1.0
Entamoeba	0 to 4	7.0	7.3	1.0
hystolitica	5 to 14	16.4	12.0	1.3
(Protozoon)	>15	16.0	13.8	1.2

Helminth eggs (but not the worms) are microscopic particles that are resistant to conventional means of disinfection such as chlorination, UV-Light and ozone (Jiménez, 2007). Compared to bacteria, only a very low infective dose (1–10 eggs instead of thousands or millions of bacteria) is needed to cause the disease. Helminth eggs (Figure 10) are considered the most resistant structure in the wastewater and sludge treatment field (Ayres et al., 1992). As a result, they are normally removed from wastewater and inactivated or destroyed in sludge. Because of their size (20–80 μm; Ayres et al., 1992) and density (1.036–1.238; Mara, 2003), ova can be efficiently removed from wastewater by sedimentation (Mara, 2003), coagulation–flocculation (Jiménez and Chavez, 1997 and 2002) or with sand filtration (Landa et al., 1997). In sludge they can be inactivated with lime (Méndez et al., 2002) or using high temperature (Feachem et al., 1983).

3.3 Wastewater treatment needs

In 1993, the government launched a project to treat wastewater, but farmers protested, requesting two things: first, to continue receiving the same amount of water; and

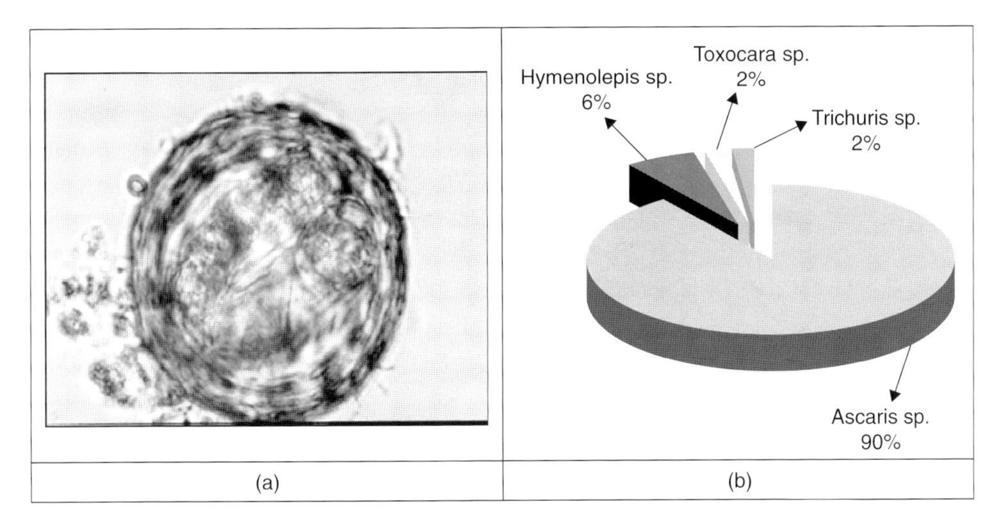

Figure 10 **(a)** *Ascaris sp.* egg and **(b)** Frequency of helminth ova genus found in Mexico City's wastewater (See also colour plate 16)

Source: Jiménez et al., 2001

second, to receive it with the 'substance' that increased crop yields. Subsequently research found that 'the substance' was the organic matter, nitrogen and phosphorus content in wastewater that fertilized the local soil. This latter request posed a serious problem because to comply with legislation precisely those compounds had to be removed. To solve this situation, the federal government first had to recognize that to reuse wastewater advantageously for agricultural irrigation, parameters considered as pollutants were not identical, and that to control the actual health risks observed helminth ova needed to be regulated[6]. This approach led to the development of a new regulation framework, the NOM-001 SEMARNAT 1996 that set a value of <1 helminth ova/L and <1,000 fecal coliforms MPN/100 mL to irrigate any kind of crops; and <5 helminth ova/L with the same fecal limit to irrigate crops used after industrialization or ingested after being cooked. No limits were established for the organic matter or the phosphorus content, while for nitrogen a limit of 40 mgN/L was defined (Jiménez, 2005).

Wastewater treatment objectives were then defined differently from the tradition in the sanitary engineering field: the helminth ova and the fecal coliform content needed to be reduced while the BOD (biochemical oxygen demand) and phosphorus had to be preserved and the nitrogen partially removed. A new set of treatment processes were considered. One of the technologies selected was an APT (advanced primary treatment), which is a low dose coagulation–flocculation process coupled with a high rate settler. These processes entail one-third of the cost of a conventional activated sludge process to meet the treatment needs and norms with a very high level of confidence (Jiménez, 1996; Jiménez and Chávez 1997 and 2002). Although for political reasons, new wastewater treatment plants for Mexico City have not been built, since the release of the NOM-001

[6]This was important not only for the Tula Valley situation but for all the country where two-thirds of the municipal wastewater produced is used for agricultural irrigation

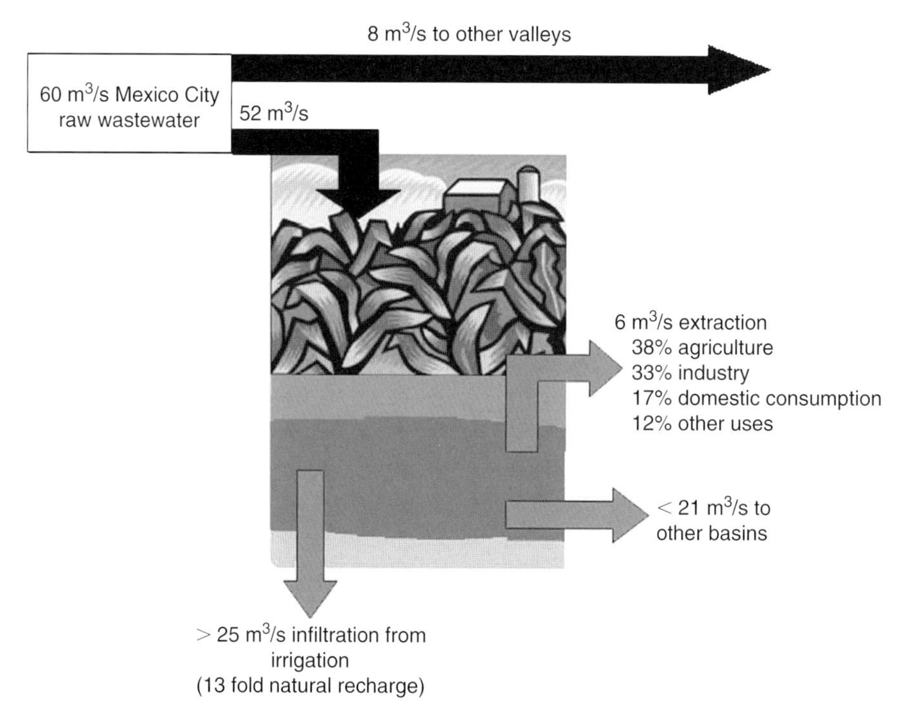

Figure 11 Water balance in the Tula Valley (See also colour plate 17)

SEMARNAT-1996. Ten wastewater treatment plants, treating 10.8 m^3/s of wastewater, have been built in Mexico using this new philosophy to irrigate agricultural fields. The process has been selected following bids open to any kind of technology.

Political problems are the result of a constant change of functionaries since 1996 in the Federal District and the Mexico State governments belonging to different political parties. Each time a new team arrives the project is invariably stopped for review and the same questions are asked: Why install high capacity wastewater treatment plants to produce water that will be used outside the Mexico Valley instead of reusing it on site? Why should the Federal District and the Mexico State governments pay to treat the water that is being used by the Hidalgo State for its development? and Why pay to treat wastewater that is appreciated by farmers? The answers come not only from technical and legal fields, but also the ethical domain, which is beyond the scope of this text.

3.4 Secondary effects of the use of wastewater for irrigation

Given that, in the Tula Valley, wastewater is transported inside hundreds of unlined channels, stored in dams and applied to a permeable soil using high irrigation rates (1.5–2.2 m/ha.yr) to wash out salts, the aquifer undergoes a significant recharge (Figure 11). In 1997, the British Geological Survey and the National Water Commission (1998) estimated that at least 25 m^3/s of wastewater were being infiltrated into the ground, representing more than 13 times the natural recharge (Jiménez and Chavez,

2004). As a result, the Tula River flow (partially fed from the aquifer) increased from 1.6 m^3/s to more than 12.7 m^3/s between 1945 and 1995, and the water table rose from being 50 m below the ground level in 1940 to form artesian wells with flows varying from 40 to 600 L/s in 1964. All these new sources of water were used to supply 500,000 inhabitants and diversify economic activities. A total of 6 m^3/s is used for agricultural irrigation (38%), industry (33%), human consumption (17%), and other uses (12%), according to Jiménez and Chavez (2004).

3.4.1 Groundwater quality

Groundwater has always been used for human consumption in the Tula Valley. Water potabilization processes were selected before the origin of the water was known. For this reason, water projects were not perceived as reuse ones and only chlorination is applied to treat water. In 1938, a change in the water quality of wells began to be noticed. And when, in 1995, it was officially acknowledged[7] that infiltrated wastewater was the origin of the new water, several studies to assess the water quality began, only this time considering it as a non-conventional water source. First, Jiménez and Chávez (2004) evaluated the characteristics of the wastewater entering the Tula Valley and compared it with that of the groundwater. More than 280 parameters (microbial, organoleptic, physical, inorganic, metal, organic compound and one toxicity test) were measured. The difference between the water quality in the aquifer formed (a 'natural reclaimed' water) and the wastewater is presented in Table 6 for some parameters. It is evident that during wastewater transportation, its use in irrigation and its infiltration through soil, it is naturally cleaned by different phenomena such as photolysis, desorption, adsorption, biodegradation and precipitation. However, it is also evident that water contained in the aquifer has a higher salt content.

Next, the quality of the drinking water was assessed, at 34 sites supplying 83% of the population, by four different laboratories working in parallel and measuring more than 220 parameters (Jiménez et al., 2003, and Jiménez and Chavez, 2004). Table 7 shows some results comparing them with the Mexican drinking-water standard. Of all the sites considered in the study, 21 exceeded the limit of total dissolved solids, 13 exceeded the limit of nitrates and fecal coliforms, 11 exceeded the limit for sodium and total hardness, 6 exceeded the limit for sulphates, 4 exceeded the limit for barium, and 1 exceeded the limits for cadmium copper, nitrites and zinc. Microbial evaluation showed a systematic presence of fecal coliforms in deficiently built wells or located inside the irrigated land, while helminth eggs and enteroviruses were never found. The most common pesticides used in the valley (atrazina, carbofurane and 2,4-D) were not found either. Nevertheless, unidentified compounds were detected in chromatograms. To identify these compounds, a second evaluation at 6 sites displaying the highest TOC (total organic carbon) and COD (chemical oxygen demand) values was performed to analyze 246 semivolatile organic compounds using the EPA SW-8270 method. Once again, none of them was found to be above the detection limit. Finally, acute toxicity tests using the Microtox® method (*Photobacterium phosphoreum*) were performed across all of the sites, with negative results. Thus, it was concluded that the water supply was

[7] But not publicly.

Table 6 Difference between Mexico City's wastewater and the Tula Valley groundwater quality in percentages, unless otherwise indicated (with information from Jiménez and Chavez (2004))

Parameter in mg/L unless indicated	*Wastewater*	*Site 1*	*Site 2*	*Site 3*
Fecal coliforms, MPN/100 mL	10^{07}–10^{10}	**8–9***	**9–10***	**7–10***
Entamoeba histolytica, No./L	0–1.5	**100**	**100**	**100**
Helminth ova, eggs/L	20–120	**100**	**100**	**100**
Salmonella, UFC/mL	0 – positive	**100**	**100**	
Shigella, UFC/mL	0 – positive	**100**	**100**	**100**
Turbidity, NTU	100–249	**99**	**99**	**99**
Total suspended solids	83–153	**97**	**98**	**97**
Total organic carbon	35–188	**84**	**75**	**90**
COD (soluble)	274–276	**97**	**97**	**96**
BOD (soluble)	166–167	**98**	**98**	**98**
Aluminum	1.3–5.5	**98**	**96**	**98**
Arsenic	ND–0.008	**71**	**56**	**82**
Copper	0.05–0.07	**77**	**67**	**82**
Total chromium	ND–0.04	**90**	**91**	**90**
Iron	1–1.2	**96**	**86**	**92**
Manganese	0.03–0.2	**95**	**88**	**95**
Lead	0.09–0.1	**78**	**78**	**84**
Bore	1–1.2	**49**	**41**	**82**
Cyanides	0.005–0.01	**13**	**33**	**17**
Fluorides	0.7–4	**74.**	**53**	**86**
Phosphorus	2.7–3	**95**	**93**	**93**
Sulphides	3–3.5	**65**	**70**	**50**
o – Xylene, µg/L	3.8–4	**100**	**100**	**100**
Ethyl benzene	1.2	**100**	**100**	**100**
m – Xylene, µg/L	9.2	**100**	**100**	**100**
p. cresol, µg/L	46.5	**100**	**100**	**100**
Chloroform, µg/L	0.2–0.8	**100**	**100**	**100**
Tetrachloroetylene, µg/L	2	**100**	**100**	**100**
Total hardness, mg $CaCO_3$/L	210–220	**−50**	**−109**	**−128**
Bicarbonates, mg $CaCO_3$/L	485	**−21**	**−12**	**−18**
Sodium	198–206	**13**	**17**	**−7**
Calcium	41–445	**−82**	**−71**	**−156**
Magnesium	24–29	**−13**	**−140**	**−76**
Chlorides	155–248	**26**	**11**	**−31**
Redox potential, mV	−16	**−215**	**−173**	**−222**
Conductivity, µmhos/cm	1437–1689	**−3.4**	**−11.3**	**−25.7**
Redox potential, mV	−16	**−215**	**−173**	**−222**
Total Nitrogen	37–38	**96**	**96**	**96**
Ammoniacal nitrogen	24–32	**97**	**100**	**100**
Nitrates	ND–1	**−2785**	**−2007**	**−2107**
Nitrites	ND–0.001	**−741**	**−521**	**−1091**

* log units
ND = Not detected

reasonably safe, but that chlorination was not the best option for disinfection due to the high values of total organic carbon content in some sites (>2 mg/L).

A third, and recent, study (Jiménez et al., 2006) was performed using water from one of the oldest springs in the Tula Valley. Using a more sensitive method to detect

Table 7 Drinking-water sources quality in the Tula Valley (Jiménez et al., 2003)

Parameter, mg/L unless indicated	*Drinking norm*	*Spring 1*	*Spring 2*	*Well 1*	*Well 2*	*Well 3*	*Well 4*	*Well 5*
Population		4,000	72,413	4,000	25,975	34,003	5,402	5,959
Volume, L/s		100	330	151	50	33	22.5	31
Fecal coliforms, MPN/100 mL	0	0	544	0	0	36	6	0
Helminth ova, ova/L	NC	0	0	0	0	0	0	0
Enteric viruses, PFU/100 mL	NC	ND	ND	ND	ND	ND	ND	ND
COD	—	17	158	0	5	48	0	0
TOC	—	0	80	0	0	0	0	0
MBAS	0.5	0.144	ND	0.112	<0.001	<0.001	<0.001	<0.001
Total hardness	500	571	465	562	250	382	340	580
Total dissolved solids	1,000	1,609	1,142	1,278	604	1,070	768	1,186
Fluorides	1.5	0.3	0.8	0.04	0.5	0.53	0.4	1.4
Nitrates	10	29	13	14	5.6	14	9	6.1
Nitrites	0.05	0.0024	0.005	0.003	0.036	0.002	0.009	<0.000001
Sulphates	400	245	167	128	74	131	60	270
Aluminum	0.2	<0.0005	ND	0.005	0.002	<0.0005	<0.0005	<0.0005
Barium	0.7	ND	ND	0.1	ND	ND	1.8	ND
Cadmium	0.005	<0.00001	0.0003	0.00007	<0.003	<0.00045	<0.0003	<0.001
Copper	2	0.008	0.004	0.006	0.004	0.002	0.002	0.002
Chrome	0.05	0.002	0.002	0.0002	0.005	0.006	0.001	0.007
Iron	0.3	0.02	0.05	0.054	0.02	0.02	<0.02	
Lead	0.025	0.0003	0.0006	<0.0002	<0.0002	0.003	<0.0002	0.0002
Sodium	200	363	210	224	95	200	56	185
Zinc	5	<0.00127	ND	2.7	0.007	0.0013	115	0.001

NC: Not considered
ND: Not detected

organic compounds (GC method 6850/6890 and EPA 590), two endocrine disrupters were detected: the Di-2 ethyl-hexyl phtalate (5.4 μg/L) and the Butyl benzyl phtalate (3.3 66 μg/L). For the first one, the US-EPA (1991) set a maximum limit of 6 μg/L. For this reason, as well as the presence of organic compounds either inducing the formation of organochlorides during chlorination or being non-identified compounds, the use of potabilization methods including membranes was recommended rather than just chlorinating the water.

3.4.2 Health effects due to the water supply in the Tula Valley

According to health experts, the main short-term risks associated with the use of a non-conventional water source relate to the presence of *Vibrio cholerae* NO-01 and other pathogens due to the presence of fecal coliforms in a content greater than

2,000 MPN/100 mL (Downs et al., 2000), but so far they appear to be been reasonably controlled by chlorination and no massive outbreaks have been reported (Jiménez et al., 2004c). In contrast, high nitrate and nitrite values (up to 29 mg N/L), exceeding 3–4 times the drinking norm, are a concern. Nevertheless, no methemoglobinemia in infants has been reported, which concurs with a recent publication prepared on behalf of the World Health Organization by Fewtrell (2004). This reference highlights that, although it had previously been accepted that consumption of drinking water with a high nitrate content causes methahemoglobinemia in infants, it now appears that nitrates may be only one of a number of co-factors that play a sometimes complex role in causing the disease. Therefore, it is inappropriate to link illness rates with drinking-water nitrate levels.

Concerning metals, their content does not exceed limits established in the drinking-water norm except in a few specific cases, and epidemiological data does not report problems. In the past, lead was thought to be a problem because a lead blood content of 7.8 μg/dL ± 4.66 μg/dL (ranging from 1.2 to 36.7 μg/dL) was found with 20% of the local population having a lead content above the threshold of 10 μg/dL (Cifuentes et al., 2000). However, detailed research showed that blood lead content was associated with the preparation of food on glazed ceramic pots, not the use or infiltration of wastewater. In addition, calcium ingestion through 'tortillas', a typical corn staple food, was found to have a protective effect against all metals. Concerning cadmium concentration, hair samples of the Tula valley population had 0.04 μg/g with a 95% confidence rate, and was not significantly different to the one found in a control group living in an area where no wastewater is used to irrigate.

Unfortunately, studies have not yet been performed to assess health risks from the recently detected organic compounds.

3.4.3 Effects on ecosystems

Besides forming a new water source, wastewater recharge has completely modified the ecology of the region (Jiménez et al., 2004c). From being a semi-desert area, the Tula Valley now has several springs and even wetlands with flora and fauna species that did not previously exist (Figure 12).

4 OPTIONS FOR THE INTEGRAL MANAGEMENT OF WATER IN THE MEXICO VALLEY

Due to the complexity of the water problem in Mexico City, a Metropolitan Water authority with the participation of the different political regions, sectors and levels of government (federal, regional and local) should be created for integrated urban water management. To be effective this Water Authority needs to have the capacity to make all decisions related to the management of water and have a sufficient budget for its operation. The main task of the Water Authority would be to elaborate a short- and long-term Integrated Water Resources Management Programme, which should consider not only the technical aspects but also the social and economical ones. Some of the activities to consider in this programme are presented below:

- Controlling soil subsidence – Several actions need to be put in place to control soil subsidence, and they include reorganization of the use of soil in the Valley of Mexico,

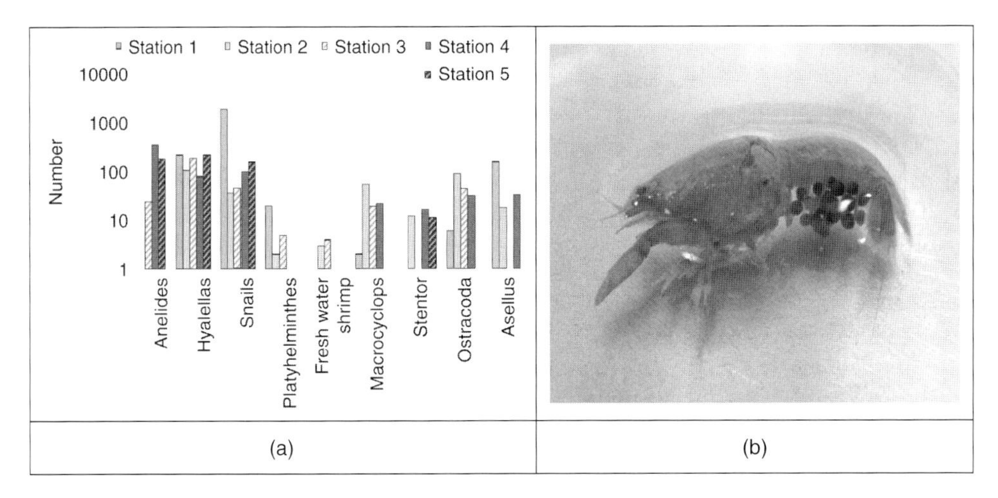

Figure 12 **(a)** Biota found in Tezontepec, a spring that appeared 30 years ago and **(b)** photograph of a Mexican shrimp named 'acocil', considered an indicator of unpolluted water (See also colour plate 18)

the increase of natural recharge, the reduction of groundwater over-exploitation and the re-injection of highly treated wastewater to the subsoil.

- Protect groundwater quality – Mexico City relies on its aquifer for water supply. Looking at this dependence it is inexplicable why almost no effort exists to protect its quality and to stop the alarmingly increase in pollution. A programme to control non-point source pollution needs to be put in place as well as the construction of sewerage infrastructure in the southern part of the city.
- Leak control – By reducing leaks in the water distribution system, groundwater over-exploitation could be reduced and the saved water could be used to meet the present water deficit. A reduction from the actual level of 40% to 20% by the year 2050 would recover 20 m^3/s.
- Implementation of an aggressive and innovative wastewater reuse programme – Mexico City is in a situation that has no precedent worldwide; therefore an integral water management programme should consider an aggressive and innovative wastewater reuse programme to fulfil needs not covered thus far. Details of this programme are given below.
- Perform innovative and comprehensive educational programmes – To obtain the support of people in implementing activities considered in the Integrated Water Management programme, it is important to explain Mexico City's water problem to the people. And, due to their complexity, this will not be an easy task. Educational programmes need to address not only society in general, but also targeted segments such as politicians, bureaucrats, industry people, academic researchers and NGOs.
- Improvement of economic tools – Water commercialization is performed deficiently by both the Federal District and the municipalities from the State of Mexico. Mexico City commercializes water with partial private participation while the municipalities of the state of Mexico perform it independently of each other and

through public participation. For the Federal District, the estimated total number of users is 2.1 million, 88% of which are registered and 60% have a meter. Around 27% (0.57 million) pay a fixed water tariff while 60% pay for the amount of water consumed. The total commercial efficiency of the system (considering billing efficiency and the amount of people that really pay) is 68%. The price of water in Mexico City is highly subsidized. Water costs at least 2 US$/m^3 while the price is 0.2 US$/m^3. Unfortunately, for political reasons water prices are not raised. The need to subsidize the water service coupled with a lack of political will to invest in the water sector has caused a serious deterioration in the related infrastructure. This is coupled with society's perception that people pay for a poor quality service. Improving economic tools together with proper communication campaigns could improve the management of water. At the same time, it should be born in mind that in setting better tariffs the enormous differences between the high and low-income segments should be considered.

- Rainwater harvesting – Mexico City has a significant pluvial precipitation representing 12 m^3/s. Therefore, several NGOs are promoting rainwater harvesting as a water supply option. Families with an important catchment area and economic resources to build the storage and water treatment facilities required can obtain 30–50% of their annual water demand this way. Unfortunately, for most of the people in Mexico City this is not a viable solution because the cost of water is very high (10 US$/m^3) and there is no possibility to build big facilities to reduce the cost because of the population density. Rainwater harvesting as a municipal service is considered as a feasible option for collecting less than 1 m^3/s, a volume that is far from the actual water need. One concern about the use of rainwater for human consumption is its water quality, since a wide variety of high organic pollutants has been reported in the polluted air of Mexico City.
- Implementation of professional public participation programmes – Mexico and Mexico City governments need to learn how to develop projects with public participation. To date this has not been done and incipient efforts have been empirical and unsuccessful. To successfully implement the Integral Water Resources Management programme, public support is needed to sustain projects independently of political parties and government changes. For that, the government needs to understand that there are professional tools for public participation.

4.1 Elements for the wastewater reuse programme

A water reuse programme for Mexico City needs to consider three sectors: agriculture, industry and local government.

Agricultural reuse – It is commonly believed that a megacity like Mexico City contains only urban areas. However, there are large agricultural areas within the city. They are located to the north (Tultitlán, Ecatepec, Jaltenco, Nextlalpan, Melchor Ocampo), north-west (Teoloyucan), west (Chalco) and south (Xochimilco). In total, these areas use 14 m^3/s of water pumped directly from the Valley of Mexico aquifer (25% of the total amount of the water extracted from groundwater and 18% of the fresh water demand). Without significantly reducing the amount of wastewater used in the Tula Valley it would be feasible to treat wastewater on site to reuse it in efficient irrigation systems. For such a project, care must be taken to exchange groundwater for reclaimed wastewater without reducing farmers' incomes. Such a programme will certainly need stakeholder involvement.

Industrial reuse – Even though 40% of national industrial activity is registered in Mexico City, this in fact refers to the corporate offices of industries located in other parts of the country, industries with low water consumption (for instance clothes manufacturing) or industries already reusing wastewater. The reason is simple, before the government initiated industrial reuse programmes, a lack of water forced industries to recycle and reuse water and to move their production facilities to other parts of the country. In addition to the already mentioned 44 private facilities reclaiming water, there are two municipal wastewater treatment plants (Lechería and Aragon) that are privately operated and are supposed to sell reclaimed water to industries. These plants sell barely 0.99 m^3/s of treated wastewater[8] because the cost of the reclaimed water is higher (around 0.7 US$/$m^3$) than the public network water (O.2 US$/$m^3$). If this problem was addressed by the government, the reclaimed water market in industry would increase to 2 m^3/s (0.5 m^3/s in the Federal District and 1.5 m^3/s in the State of Mexico), an amount that represents only 3% of the total wastewater generated, or 2% of the fresh water demand. Thus, while large, the potential to reuse reclaimed water is limited.

Municipal reuse – Because water professionals in the government of Mexico City have always been conscious of the lack of water, municipal reuse is already a common practice. All options for non-potable reuse wastewater have already been put into practice; now the need is to reuse wastewater for human consumption. This can be performed in two ways: by highly treating wastewater locally; and, by importing the excess water from the Tula Valley aquifer formed from the infiltration of Mexico City's wastewater.

Wastewater treatment on site – Considering the experience of Windhoek, Namibia (Van der Merwe et al., in press) this is a feasible option. For Mexico City, however, it has the drawback of reducing the amount of wastewater sent to the Tula Valley by around 14.5 m^3/s to render only 10 m^3/s of supply[9] which is less than the amount needed and implies a sophisticated and expensive treatment process. Furthermore, this option seems less likely to be accepted by the public. An advantage is that on-site wastewater treatment eliminates transportation costs.

Recovering the water sent to the Tula Valley – As mentioned, wastewater recharge is at least 25 m^3/s (BGS-CNA, 1998) in the Tula Valley and wastewater is being highly treated through its transportation in open channels and used for irrigation and infiltration into the soil. This option has three drawbacks. First, independent of the present natural treatment of the wastewater, it is necessary to treat Mexico City's wastewater to fulfil the norms and control the health risks created by the use of the wastewater, but also maintain the soil's clean-up capacity which in some places seems to be overloaded (Jiménez et al., 2006). Second, there is the cost of transporting water back to the city 100 km away and at a 150 m height difference (Jiménez et al., 1997). Third, a political negotiation is needed because, although by law groundwater belongs to the nation, Hidalgo's population thinks that is why they should return 'their' water to Mexico City after having cleaned it. Advantages of this option are: the

[8] Total capacity of reclaiming plants is 1.8 m^3/s.
[9] Considering a 70% recovery of the injected reclaimed wastewater.

Table 8 Cost comparison of different new water source options for Mexico City (Jiménez and Chavez, 2004)

Source	Flow m^3/s	Pumping US$/$m^3$	Investment US$/$m^3$	Treatment cost US$/$m^3$	Total cost US$/$m^3$
Temascaltepec	5	0.30	0.27	0.016	0.58
Amacuzac	13.5	0.40	0.44	0.016	0.86
Tecolutla I	9.8	0.48	0.36	0.016	0.85
Potabilization of raw wastewater	3	0	0	0.95	0.95
On site					
Tula Valley aquifer	6	0.09	0.035	0.50	0.73

Tula Valley will continue to use the same amount of water; the extraction of groundwater will help to control salinization and flooding problems observed in the lower agricultural fields of the Tula Valley due to the rise in the groundwater level;[10] and it is cheaper than potabilizing the wastewater on site.

Table 8 compares the cost of different alternative water sources for Mexico City involving reuse and freshwater sources. The cheapest option is to import water from Temascaltepec; however, this is not viable due to the resistance of the local community to giving up its water to Mexico City.

Independently of the option selected, local governments (Federal District and the municipalities of the State of Mexico) must be aware that:

- Prior to implementing any of the reuse programmes described, it is necessary to segregate or treat industrial discharges, even though they are not volume important.
- The direct reuse of wastewater for human consumption is a controversial option (Asano, 1998). Any decision taken will need to be based on technical, social, economic and scientific studies that imply an elevated cost and several years to perform. It is estimated that at least 4–6 million US$ are needed to obtain all the scientific, technical and practical information needed in 5–7 years.

For both options presented, it is striking that although they are frequently cited as a priority and a next step for the city by politicians, functionaries and even academics, these 'solutions' have never been submitted to public consultation, although rejection could stop the project. In preparation for the project, the Federal District government published a norm (2004) to reuse wastewater for human consumption (parameters considered are presented in Appendix 1) but the Federal Government blocked its application for political reasons.

5 CONCLUSIONS

Mexico City is undoubtedly experiencing a very challenging situation concerning its water supply and wastewater disposal system which may have no precedents in other

[10] The rise in the water table is actually causing a loss of 0.95 m^3/s of water through evaporation.

parts of the world. Nevertheless, it is highly likely that at least some of the problems discussed are already being experienced elsewhere, hence the need to create awareness about the importance of managing urban water integrally, particularly in megacities.

BIBLIOGRAPHY

Asano, T. 1998. *Wastewater Reclamation and Reuse*. Water Quality Management Library. Vol. 10, Ed. Technomic Publishing Company, United States.

Ayres, R., Alabaster, G., Mara, D. and Lee D. 1992. A Design Equation for Human Intestinal Nematode Egg Removal in Waste Stabilization Ponds, *Water Research*, Vol. 26, No. 6, pp. 863–865.

Bellia, S., Cusimano, G., González, M.T., Rodríguez, R.C. and Giunta, G. 1992. *The Mexico Valley, Preliminary Considerations of the Geological and Hydrogeological Risks*. Quaderni Instituto Italo-Latino Americano. Serie Scienza 3. Roma, Italia: 96 pp. [In Spanish].

BGS-CNA (British Geological Survey and National Water Commission). 1998. Impact of Wastewater Reuse on Groundwater in the Mezquital Valley, Hidalgo State, Mexico. *Final Technical Report WC/98/42*.

Capella, A. 2006. Personal communication.

Cifuentes, E., Blumenthal, J., Ruiz-Palacios, G. and Beneth, S. 1992. Health Impact Evaluation of Wastewater in Mexico, *Public Health Revue*, Vol. 19, pp. 243–250.

Cifuentes, E., Suárez, L., Solano, M. and Santos, R. 2002. Diarrheal Diseases in Children from Water Reclamation Site, Mexico City. *Environmental Health Perspectives*, Vol. 110, No. 10, pp. 619–624.

Cifuentes, E., Villanueva, J. and Sanin, H. 2000. Predictors of Blood Lead Levels in Agricultural Villages Practicing Wastewater Irrigation in Central México. *International Journal of Occupational and Environmental Health*, Vol. 6, No. 3, pp. 177–182.

DDF. 1985. *Geohydrological Activities in the Valley of Mexico*. General Department of Construction and Hydraulic Operation. Contract 7-33-1-0403. México, D.F. [In Spanish].

DGCOH (General Department of Construction and Hydraulic Operation). 1998. The Water Situation in the Federal District, Gobierno del DF. [In Spanish].

DOF, Official Federation Diary. 2000. Modification to the Mexican Official Norm NOM-127-SSA1-1994. *Water for Use and Human Consumption. Allowable Quality Limits and Treatment to Potabilize Water*. October 20, 1–8 [In Spanish].

Domínguez, R. 2006. Personal communication.

Domínguez-Mora, R., Jiménez-Cisneros, B., Carrizosa-Elizondo, E. and Cisneros-Iturbe, L. 2005. *Water Supply and Flood Control in the Mexico Valley*. No. 06-cd-03-10-0274-1-05, Project 5341, Engineering Institute UNAM [In Spanish].

Downs, T., Cifuentes, E., Ruth, E. and Suffet, I. 2000. Effectiveness of Natural Treatment in a Wastewater Irrigation District of the Mexico City Region: A Synoptic Field Survey. *Water Environment Research*, Vol. 72, No. 1, pp. 4–21.

Ezcurra, E., Mazari, M., Pisanty, P. and Guillermo, A. 2006. *The Mexico Valley*. Fondo de Cultura Económica Ed., Serie Ciencia y Tecnología, México, 285 pp. [In Spanish].

Feachem, R., Bradley, D., Garelick, H. and Mara, D. 1983. *Sanitation and Disease: Health Aspects of Excreta and Wastewater Management*. John Wiley and Sons, New York.

Fewtrell, L. 2004. Drinking-Water Nitrate and Methemoglobinemia. Global Burden of Disease: A Discussion, *Environmental Health Perspectives*, Vol. 112, No. 14, pp. 1371–1374.

Gaceta Oficial del Distrito Federal, Official Journal of the Federal District. 2004. *Environmental Norm for the Federal District* NADF-003-2002 to establish the conditions and requirements to directly recharge treated wastewater to the Federal District Aquifer, 26 March 2003 [In Spanish].

Garza, G. and Chiapetto, C. 2000. Mexico City into the National Urban System. Garza, G. (ed.) *Mexico City towards the End of the Second Millennium*. Gobierno del Distrito Federal and El Colegio de México, Mexico, pp. 765 [In Spanish].

GDF, Federal District Government. 1999. *Compendium DGCOH 1999*. Gobierno del Distrito Federal, Secretaría de Obras y Servicios. Dirección General de Construcción y Operación Hidráulica Ed, pp. 82 [In Spanish].

Guerrero, L., Calva, J.J., Morrow, A.L., Velázquez, R., Tuz-Dzib, F., López-Vidal, Y., Ortega, H., Arroyo, H., Cleary, T.G., Pickering, L.K. and Ruiz-Palacios, G., 1994. Asymptomatic, Shigella Infections in a Cohort of Mexican Children Younger than Two Years of Age, *Pediatric Infectious Diseases Journal*, Vol. 13, pp. 596–602.

Guillermo, A. 2000. Geographical Localization of the Mexico Basin. Garza, G. (ed.) *Mexico City Towards the End of the Second Millennium*. Gobierno del Distrito Federal and El Colegio de México Ed., Mexico, pp. 765 [In Spanish].

INEGI, National Institute of Statistics, Geography and Informatics. 2005. www.inegi.gob.mx/est/default.asp?c=119.

Jiménez, B. 2005. Treatment Technology and Standards for Agricultural Wastewater Reuse: A Case Study in Mexico, *Irrigation and Drainage*, Vol. 54, No. 1, pp. 23–35.

Jiménez, B. 2007. Helminth Ova Removal from Wastewater for Agriculture and Aquaculture Reuse, *Water Science and Technology*, Vol. 55, No. 1–2, pp. 485–493.

Jiménez, B. and Asano, T. In press. *International Wastewater Reuse Survey*, IWAP Ed.

Jiménez, B. and Chávez, A. 1997. Treatment of Mexico City Wastewater for Irrigation Purposes, *Environmental Technology*, Vol. 18, pp. 721–730.

Jiménez, B. and Chávez, A. 2002. Low Cost Technology for Reliable Use of Mexico City's Wastewater for Agricultural Irrigation, *Technology*, Vol. 9, No. 1–2, pp. 95–108.

Jiménez, B. and Chávez, A. 2004. Quality Assessment of an Aquifer Recharged with Wastewater for its Potential Use as Drinking Source: "El Mezquital Valley" case, *Water Science and Technology*, Vol. 50, No. 2, pp. 269–273.

Jiménez, B., Barrios, J.E. and Cruickshank, C. 2003. Evaluation of a Wastewater Recharged Aquifer as a Source of Water Supply. *11th Biennial Symposium of Groundwater Recharge*. Arizona Hydrological Society, Arizona Department of Water Resources, Salt River Project, US Water Conservation Laboratory of USDA-ARS,. 5–7 June Tempe, Arizona, USA.

Jiménez, B., Chávez, A. and Capella, A. 1997. Wastewater in the Valley of Mexico and its Reuse. *70th Annual Conference and Exposition*, 7, 2/32, 311-320, ISBN-1-57278-111-3, IAWQ. Chicago, Illinois, Estados Unidos.

Jiménez, B, Murillo, R. and Chavez, A. 2006. *Detection of Toxic and Endocrine Disrupter in Well of the Tula Valley*. Report No. 3333, Engineering Institute Agreement CONACYT SEMARNAT-2002-01-0519 [In Spanish].

Jiménez, B., Siebe, C. and Cifuentes, E. 2004c. Intentional and Non Intentional Reuse of Water in the Tula Valley. In *The Water from the Academia Perspective*. Academia Mexicana de Ciencias, Mexico, pp. 33–55. [In Spanish].

Jiménez, B., Chávez, A., Maya, C. and Jardines, L. 2001. The Removal of a Diversity of Microorganisms in Different Stages of Wastewater Treatment, *Water Science and Technology*, Vol. 43, No. 10, pp. 155–162.

Jiménez, B., Mazari, M., Cifuentes, E. and Domínguez, R. 2004a. The Water in the Mexico Valley. In *The Water from the Academia's Perspective*. Academia Mexicana de Ciencias, Mexico pp. 15–32 [In Spanish].

Jiménez, B., Maya, C., Lucario, S., Chavez, A. and Becerril, E. 2004b. *Evaluation of Tap Water in a Suburb of Mexico City*. XIV FEMISCA National Congress Mazatlán, Sinaloa, May. [In Spanish].

Jiménez-Cisneros, B. 1996. Wastewater Reuse to Increase Soil Productivity, *Water Science and Technology*, Vol. 32, No. 12, pp. 173–180.

Landa, H., Capella, A. and Jiménez, B. 1997. Particle Size Distribution in an Effluent from an Advanced Primary Treatment and its Removal during Filtration, *Water Science and Technology*, Vol. 36, No. 4, pp. 159–165.

LeBaron, C.W., Lew, J., Glass, G.I., Weber, J.M., Ruiz-Palacios, G. and the Rotavirus Study Group. 1990. Annual Rotavirus Epidemic Patterns in North America, *The Journal of the American Medical Association*, Vol. 264, No. 8, pp. 983–988.

Lesser, J.M., Sánchez, F. and González, D. 1986. Hydrochemistry of the Aquifer of Mexico City, *Hydraulic Engineering*, Sept–Dec, pp. 64–77 [In Spanish].

López-Vidal, Y., Calva, J.J., Trujillo, A., Ponce de León, A., Ramos, A., Svennerholm, A.M. and Ruiz-Palacios, G., 1990. Enterotoxins and Adhesins of Enterotoxigenic *Escherichia coli*: Are they a Risk Factor for Acute Diarrhea in the Community? *The Journal of the Infectious Diseases*, Vol. 162, No. 2, pp. 442–447.

Mara, D. 2003. *Domestic Wastewater Treatment in Developing Countries*. Ed. Earth Scan, London.

Mara, D. and Cairncross, S. 1989. Guidelines for the Safe Use of Wastewater and Excreta in Agriculture and Aquaculture. World Health Organization, Geneva, pp. 185.

Mazari-Hiriart, M., López Vidal, Y., Castillo-Rojas, G., Ponce de León, S. and Cravioto, A. 2001. *Helicobacter pylori* and Other Enteric Bacteria in Freshwater Environments in Mexico City, *Archives of Medical Research*, Vol. 32, No. 5, pp. 458–467.

Mazari-Hiriart, M., López-Vidal, Y., Ponce de León, S., Calva-Mercado, J.J. and Rojo-Callejas, F. 2002. Significance of Water Quality Indicators: A Case Study in Mexico City. *Proceedings of the International Conference: Water and Wastewater, Perspectives of Developing Countries*. Indian Institute of Technology Delhi-International Water Association. New Delhi, India. December, No. 11–13, pp. 407–416.

Mazari-Hiriart, M., Torres Beristain, B. Velázquez, E., Calva, J. and Pillai, S. 1999. Bacterial and Viral Indicators of Fecal Pollution in Mexico City's Southern Aquifer, *Journal of Environmental Science Health*, Vol. A34, No. 9, pp. 1715–1735.

Méndez, J.M., Jiménez, B. and Barrios, J.A. 2002. Improved Alkaline Stabilization of Municipal Wastewater Sludge. *Water Science and Technology*, Vol. 46, No. 10, pp. 139–146.

Merino, H. 2000. Hydraulic System. Garza, G. (ed.) *Mexico City Towards the End of the Second Millennium*. Gobierno del Distrito Federal and El Colegio de México Ed., Mexico, pp. 765.

Saade Hazin, L. 1998. New Strategy in Urban Water Management in Mexico: The case of Mexico's Federal District, *Natural Resources Forum*, Vol. 22, No. 3, pp. 185–192.

SACM. 2006. Mexico City Water Works, International Tlalocan Festival. *Water and Politics. Tlalocan Proceedings*, 10–15 March, Mexico City.

Santoyo, E. and Ovando-Shelley, E. 2002. Underexcavation at the Tower of Pisa and at Mexico City's Metropolitan Cathedral. *Proc. International Workshop*, ISSMGE-Technical Committee TC36 Foundation Engineering in Difficult, Soft Soil Conditions, CD edition, Mexico City [In Spanish].

Santoyo, E., Ovando, E., Mooser, F. and León, E. 2005. *Geotechnical Syntheses of the Mexico Valley Basin*. TGC, Geotecnia, S.A. de C.V. México, pp. 171 [In Spanish].

Siebe, Ch. 1998. Nutrient Inputs to Soils and their Uptake by Alfalfa through Long-term Irrigation with Untreated Sewage Effluent in Mexico, *Soil Use and Management*, Vol. 14, pp. 119–122.

SSA. (Health Secretariat). 2005. *Morbility Yearbook*. Sistema Único para la Vigilancia Epidemiológica. Dirección General de Epidemiología. http://www.dgepi.salud.gob.mx [In Spanish].

UN. 2003. *Water for People Water for Life*. The United Nations World Water Development Report. Barcelona, UNESCO.

US EPA. 1991. *Final Drinking Water Criteria Document for Phthalic Acid Esters (PAES)*. NTIS # PB92-173442, 321.

US EPA. 2004. *Guidelines for Water Reuse.* EPA/625/R-04/108, USAID, Washington, D.C.

Van der Merwe, B., du Pisani, P., Menge, J. and König, E. In press. Water Reuse in Windhoek, Namibia: 37 years and still the Only Case of Direct Water Reuse for Human Consumption. Asano, T and Jiménez B, (eds.) *International Wastewater Reuse Survey*, IWAP.

WHO. 1995. *Drinking Water Quality Criteria*, 2nd Edition. Geneve Switzerland. OMS Ed.

WHO. 2006. *Guidelines for the Safe Use of Wastewater, Excreta and Greywater.* Volume 2, Wastewater Use in Agriculture. WHO Library Cataloguing-in-Publication Data, Geneva, pp. 213.

World Health Organization (WHO). 1989. *Health Guidelines for the Use of Wastewater in Agriculture and Aquaculture.* Technical Report Series No. 778, Geneva, pp. 69–85.

APPENDIX 1 Parameters considered in a norm proposed by the Federal District to regulate the local reinjection of reclaimed water intended for human consumption reuse

Parameter	*Maximum limit*	*Parameter*	*Maximum limit*
E. coli, fecal coliformes or Thermotolerant organisms	Absent	Nitrates, as N	10.00
Enteroviruses	Absent	Nitrites, as N	1.0
Fecal Streptococcus	Absent	Ammonical nitrogen as N	0.5
Giardia lamblia	Absent	Percloroethylene or tetrachlorethyilene	0.040
Color	15 Pt-Co units	pH	6.5–8.5
Conductivity	$<$15% higher to that of the aquifer	1,2-dibromo-3-chlore propane, in μg/L	1
Turbidity	5 NTU	2,4 *D*, in μg/L	30.00
Alum	0.2	Alachlore, in μg/L	20
Arsenic	0.025	Aldicarb, in μg/L	10
Barium	0.7	Aldrín and dieldrín, in μg/L	0.03
Benzene	0.01	Atrazine, in μg/L	2
Boron	0.3	Carbofurane, in μg/L	5
Cadmium	0.003	Chlordane, in μg/L	0.2
Total organic Carbon	1.0	DDT, in μg/L	1.00
Cyanides	0.07	Gama-HCH (lindane), in μg/L	2.00
Free residual chlorine	0.0	Heptachlore and heptha chlore exopic, in μg/L	0.03
Chlorides as Cl^-	250.00	Hexaclore benzene, in μg/L	1.00
Vinyl chloride	0.005	Metoxichlore, in μg/L	20.00
Copper	2.00	Lead	0.01
Total Chrome	0.05	ABS	0.5
1,1-Dichlore ethylene	0.030	Sodium	200.00
Total hardness as $CaCO_3$	500.00	Total dissolved solids	1000.00
Styrene	0.02	Total suspended solids	5.00
Ethyl benzene	0.3	Sulphates, as SO_4^{2-}	400.00
Phenols	0.3	Toluene	0.7
Fluorides	1.5	Total Tri halomethanes	0.20
Phosphorus as PO_4^{3-}	1.0	Triclorethylene	0.070
Polyaromatic hydrocarbons PAHs	0.0002	1,1,1-Trichlore ethane	2.0
Iron	0.3	Xylenes (three isomer)	0.5
Manganese	0.15	Zinc	5.00
Mercury	0.001		
Methyl terbuthyl ether	0.03		

Case Study II

Integrated urban water management in the Tucson, Arizona metropolitan area

Robert G. Arnold[1] and Katherine P. Arnold[2]

[1]Department of Chemical and Environmental Engineering, The University of Arizona, Tucson, Arizona, USA
[2]Malcolm Pirnie Consulting Engineers, Tucson, Arizona, USA

1 INTRODUCTION

About two-thirds of the fastest growing American cities are in portions of the country where water supply sustainability is an issue. Arizona and Nevada, both of which represent 'water is for fightin' states, are experiencing the highest rates of population and economic growth in the United States. Until recently, many western communities relied, disproportionately (relative to other geographic sectors of the country), on groundwater. The finite nature of groundwater mandates consideration of waters of increasingly impaired quality for inclusion in regional water resource portfolios. These include imported surface waters with relatively high salinity, such as the lower Colorado River; brackish groundwater; and, in virtually every major south-western municipality, reclaimed wastewater. Water supply in western cities has produced tensions related to:

- The allocation of limited resources among competing uses.
- The use-dependent degree of treatment for waters of lower initial quality.
- Sustainability objectives – essentially the degree to which future water quality and quantity should be considered when making contemporary water supply decisions.
- Economic realities – the Tucson metropolitan area and vicinity is a microcosm for water management decisions in the semi-arid south-west.

2 PHYSICAL AND INSTITUTIONAL FRAMEWORK

Prior to 2001, residents of the Tucson metropolitan area relied exclusively on groundwater to satisfy potable water demand. There was essentially no engineered effort to replenish the regional groundwater resource. Tucson's groundwater is of exceptional quality and was distributed for potable use with no treatment other than disinfection. The situation was not sustainable, however, and the water table declined by up to 1.2 metres per year in major Tucson well fields.

Over-reliance on groundwater is widespread in Arizona. After the passage of Arizona's Groundwater Management Act of 1980 (GMA), the Arizona Department of Water Resources (ADWR) established five Active Management Areas (AMAs), encompassing the State's most populous, water-stressed regions. Specific groundwater

Table 1 Comparison of regional Tucson groundwater and CAP water quality characteristics. All figures are in mg/L. Groundwater data are spatial averages that encompass considerable variability. CAP data represent water quality characteristics at the southern terminus of the CAP canal, in the vicinity of Tucson. Average values were obtained from 2006 water quality measurements

Water quality constituent (mg/L)	*Tucson water production wells*	*CAP water*
Total Dissolved Solids	259	666
Hardness (as $CaCO_3$)	119	270
Sodium	40	112
Chloride	17	104
Calcium	39	56
Magnesium	5	31
Sulphate	45	280
Alkalinity	126	98
TOC	<1	3.5

management goals were set for each AMA, reflecting local conditions. The Tucson Active Management Area (TAMA) is a so-called 'safe-yield' AMA, and must regulate the use of regional groundwater resources so that a balance of groundwater withdrawals and replenishment, or safe-yield condition, will be established by the year 2025. The TAMA also has 'assured water supply' requirements that are described later. Throughout the following case study, the TAMA is taken as the primary regional water-planning unit. When projections or analyses apply to other hydrologic units or to specific water providers, a distinction is made in the text.h

The TAMA covers 10,013 square kilometres in southern Arizona. The area encompasses two major hydrologic units, the Avra Valley Sub-basin and the northern portion of the Upper Santa Cruz Valley Sub-basin. Average annual precipitation varies from 28 centimetres at low elevations to 71 centimetres in the mountains. The primary surface water feature in the TAMA is the Santa Cruz River, which flows generally northward for about 100 kilometres within the TAMA boundaries. The river is effluent-dependent, however, and perennial flow is observed only in a 25-kilometre reach that lies north of the two regional wastewater treatment plants in Tucson. Brief periods of high flow are observed after intense rainfall.

Natural groundwater movement within the TAMA is north-west. Declines in groundwater levels due to over-pumping between 1940 and 1995 were estimated at 60 metres in parts of Tucson and 45 metres in the Green Valley/Sahuarita area. The depth to groundwater in the TAMA generally ranges from 60–200 metres and is lowest in the vicinity of the river. The average net rate of natural groundwater recharge in the TAMA due to infiltration and groundwater underflow at TAMA boundaries has been estimated at 75 million cubic metres per year ($m^3 \cdot yr^{-1}$). The figure is based on estimates of natural infiltration rates, groundwater inflow from the Santa Cruz AMA and groundwater outflow to the Pinal AMA. When averaged over the entire TAMA surface, however, this is only 0.75 centimetres per year.

The GMA mandates attainment of 'safe yield', or a balance between groundwater withdrawals and replenishment over the TAMA (taken as a single hydrologic unit), by 2025. Furthermore, land development must be preceded by a demonstration that

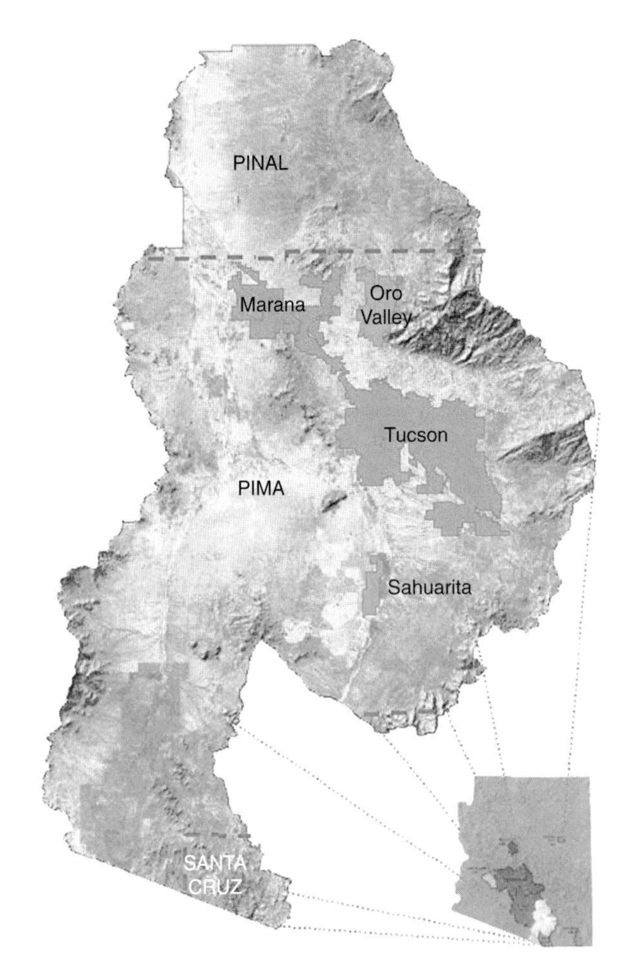

Figure 1 The Tucson Active Management Area (TAMA). The insert is provided to show the general location of the TAMA and positions of other Arizona AMAs (See also colour plate 19)

Source: Arizona Department of Water Resources

there is an 'assured water supply'. Arizona's 1995 Assured Water Supply (AWS) Rules prohibit growth that interferes with an AMA's ability to provide water of a quantity and quality necessary to satisfy regional water demand for the next 100 years while meeting safe-yield requirements.

Water demand in the TAMA has outgrown the local renewable groundwater supply. Sustained growth is possible only because Colorado River water is delivered to the Tucson area via the Central Arizona Project (CAP) canal. Expanded use of Colorado River water within the TAMA makes co-management of surface and groundwater a necessity. The Central Arizona Water Conservation District (CAWCD), also known as the Central Arizona Project (CAP), operates the CAP canal. The price of delivered water is set to reimburse the federal government for canal construction costs. The City of Tucson currently purchases CAP water from the CAWCD for \$88 per 1,000 m^3.

Table 2 Major water providers in the TAMA based on population served (1998 estimates unless otherwise indicated). For those with CAP water subcontracts, allocations are provided. Water volumes/time are in millions of cubic metres per year

Water system	Population served	Volume distributed[1]	CAP allocation
Tucson Water Department	553,040	135.6	171.4
Metropolitan Domestic Water Improvement District	36,250	10.6	10.9
Flowing Wells Irrigation District	16,160	3.51	5.37
Oro Valley Water Dept.	10,850	7.04	2.83
Green Valley Water Co.	8,125	2.54	2.34
Community Water – Green Valley	12,320		1.65
University of Arizona	15,950		
USAF – Davis Monthan AFB	8,900		
Pima County Parks	9,142		
Amphitheater School District	10,615		
Other with CAP Subcontracts			
San Xavier District – Tohono O'odham Nation			33.3
State Land Department			17.3
Schuk Toak District – Tohono O'odham Nation			13.3
Spanish Trail Water Company			3.75
Midvale Farms			1.85
Vail Water Company			0.97
Pascua Yaqui Tribe			0.62
Town of Marana	3,277	0.64	0.06

[1] 1995 data; from WRRC (1999)

The Central Arizona Groundwater Replenishment District (CAGRD) is a division of the CAWCD that is authorized to acquire water supplies for aquifer replenishment in districts served by the CAWCD. The CAGRD mechanism is used by local water providers when immediately accessible water resources do not meet AWS requirements for growth. That is, water providers can pay the CAGRD to recharge unused CAP water, compensating for a local groundwater overdraft. The offset water can be recharged anywhere in the TAMA, however, without considering local effects on groundwater levels.

The Tohono O'odham and Pascua Yaqui Indian Tribes maintain unused rights to CAP water and groundwater within the TAMA. Negotiations leading to utilization or leasing of Indian water rights could provide clarity and flexibility to safe yield and AWS determinations in the TAMA. The United States Bureau of Reclamation (USBR) represents the tribes in Indian water rights negotiations on behalf of the Department of Interior. The Bureau maintains a broad interest in regional water supply planning and the development of treatment options for Colorado River water.

The City of Tucson is by far the largest municipal water purveyor in the TAMA (Table 2). The Tucson Water Department's (Tucson Water) service area extends well beyond the city boundaries. About 40% of those served by Tucson water reside outside the city boundaries. Nevertheless, Tucson Water is a city-owned utility whose

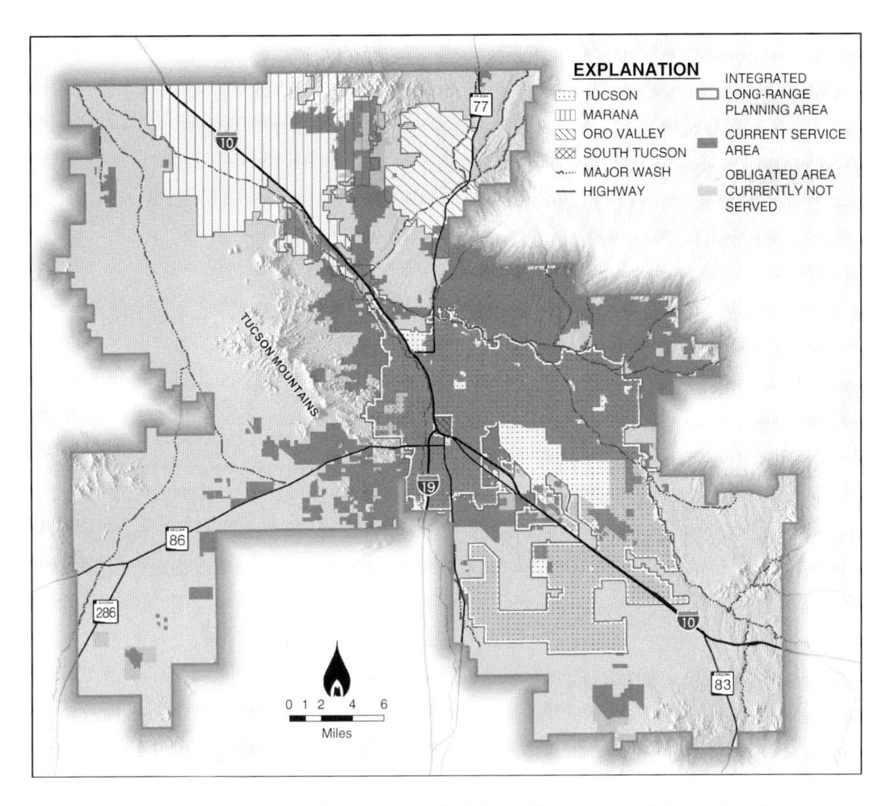

Figure 2 Boundaries of Tucson Water's current and obligated service and that of the long-range, integrated water supply planning area in the Tucson region. (Courtesy of Tucson Water)

rates are authorized by Tucson's mayor and council. Although at least 30 water companies operate wells in the Tucson area, the most significant in terms of population served are the Metropolitan Domestic Water Improvement District, Oro Valley Water Department and the Flowing Wells Irrigation District. About 22,000 individuals and businesses have their own wells.

3 TAMA DEMAND AND SUPPLY

3.1 Preliminary considerations

Some notion of safe yield is essential to understanding water use constraints in the TAMA. The renewable water supply in the TAMA is equal to the natural rate of groundwater replenishment plus the regional CAP allotment and the amount of wastewater that can be reclaimed and reused. Groundwater is naturally replaced at a rate of $75 \times 10^6\,m^3 \cdot yr^{-1}$, and while the full allotment of CAP water to TAMA subcontractors is now $266 \times 10^6\,m^3 \cdot yr^{-1}$, that figure is projected to increase to $324 \times 10^6\,m^3 \cdot yr^{-1}$ as a result of pending CAP water reallocations. Expansion of the regional water supply beyond these limits without resorting to groundwater mining or identification of a major new water resource will depend on water reuse. There is a physical limit to the sustainable water supply, even if municipal wastewater is completely reused without consideration of water quality, legal, fiscal, political or other practical constraints. In a crude way, that limit is given by:

$$Q_{\max} = \frac{Q_{new}}{1 - R} \quad (1)$$

where Q_{new} is the rate at which renewable water can be supplied, neglecting water reuse
R is the fraction of water supplied that is recovered and reused, and
Q_{max} is the maximum sustainable water supply.

The recovered water fraction (R), based on recent regional demand and wastewater treatment data, is about 0.44. Thus, the maximum sustainable rate of water supply in the TAMA is about $710 \times 10^6 m^3 \cdot yr^{-1}$, unless the recovered water fraction is increased or additional renewable water resources can be identified. In the next section, the regional water demand and supply functions are broken down in detail, and this estimate is refined.

3.2 Water demand history

The three primary water use categories are agricultural, industrial and municipal. Agricultural water demand in the TAMA declined precipitously from a maximum of $370 \times 10^6 m^3$ in 1975 to $125 \times 10^6 m^3$ in 1985, due to a reduction in regional farm acreage, but remained steady during 1985–2000. Treated wastewater provides just 3% of water used regionally for agriculture, so that virtually all agricultural demand is satisfied from groundwater.

Industrial use varied from 15 to 20% of total TAMA water demand during 1984–2000. The mining industry typically accounts for about two-thirds of industrial demand, and water use for mining fluctuates primarily with the price of copper. Turf and gravel industries account for another 20%. During 1995–2000, industrial water rights ($\sim 245 \times 10^6 m^3 \cdot yr^{-1}$) were considerably greater than the rate of industrial water use ($\sim 75 \times 10^6 m^3 \cdot yr^{-1}$) in the TAMA. Unused water rights arise in part from industrial users that ceased or reduced operations but retained their water rights. Essentially all industrial water demand is satisfied using groundwater.

Ninety percent of municipal water used in the TAMA is provided by the five largest water purveyors. Tucson Water alone distributed >75% of the municipal water served in 1995. During 1975–2000, municipal water demand in the TAMA increased from $<125 \times 10^6 m^3 \cdot yr^{-1}$ to almost $200 \times 10^6 m^3 \cdot yr$. Per capita municipal water demand varied between 666 and 742 litres per capita per day (LPCD) with essentially no time-dependent trend. Tucson Water's average municipal demand was less than 640 LPCD during 1985–1986, and the average residential demand was just 416 LPCD. The latter figure excludes commercial uses that are satisfied from the municipal distribution system (132 LPCD) and lost or unaccounted for water (68 LPCD). Single-family residential demand is already low relative to other major western cities, suggesting that additional water conservation measures can produce only modest returns.

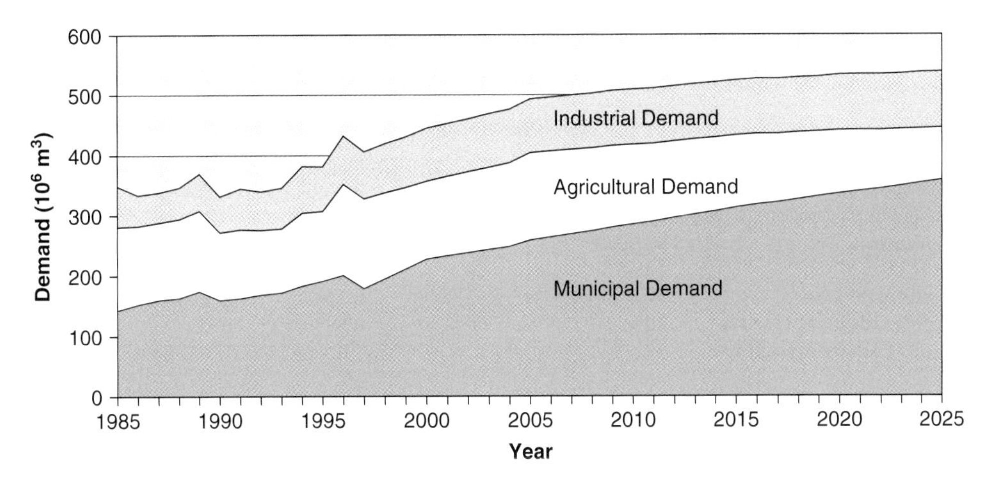

Figure 3 Historical and projected water demand in the TAMA. The top line represents total regional demand, and the shaded regions below that line correspond to the three major components of overall demand. It is evident that growth in total water demand is dominated by the municipal component

Table 3 Industrial water rights and 1995 water withdrawals, by industry, in the TAMA. All figures are in millions of cubic metres per year

User category	*Right (permit allotment)*	*1995 total water use*	*1995 groundwater use*
Mining	77.8	51	51
Turf-Related[1]	13.6	9.82	8.86
Sand and Gravel	21.2	6.38	6.38
Other Industrial	112	4.97	4.97
Power Generation	12.4	1.98	1.98
Dairies	0.26	0.09	0.09
TOTAL	**237.4**	**74.3**	**73.3**

[1]Turf irrigation from the Tucson Water reclaimed water distribution system is not included in this figure. That water is accounted for as municipal demand.

The distribution of reclaimed water for landscape irrigation was initiated in about 1990. Between 1990 and 2000, municipal use of reclaimed water increased from 13.6×10^6 to $19.6 \times 10^6 \mathrm{m}^3 \cdot \mathrm{yr}^{-1}$, and further increases are projected. However, such uses are tallied by the State outside the calculation of per capita demand. Consequently, municipal water reuse should be added to formal municipal demand projections based on population/GPCD trends.

3.3 Water supply

The physical basis of water supply planning in the TAMA is deceptively simple. Sustainability depends on a reasonable balance between water supply and demand

Table 4 Year 1995 water deliveries and population served by major water providers in the TAMA. The five water providers shown account for about 90 percent of municipal water served in the TAMA. Water use data are in millions of cubic feet per year

	Providers				
	City of Tucson	*Oro Valley*	*Marana*	*MDWID*	*FWID*
Population	597,017	22,479	3,277	40,870	14,951
Total Use	135.6	7.04	0.64	10.56	3.51
Residential Use	91.8	3.52	0.56	8.35	2.63
% of Residential Use by Single-Family Dwellings	70%	97%	89%	89%	45%

The year 1995 TAMA water budget is summarized in Table 6.

Table 5 Single-family residential (per capita) water demand in representative western cities. Year 2003 figures are from utility representatives except for Las Vegas, which was from Western Resource Advocates (2003)

Municipality	*Per Capita Municipal Use (LPCD)*
Albuquerque	360
El Paso	432
Tucson	454
Tempe	530
Phoenix	625
Las Vegas	871
Sacramento	916
Fresno	988

without long-term dependence on groundwater mining. The primary renewable water resources in the TAMA are:

- water that is naturally recharged to the regional aquifer
- CAP water
- treated wastewater that is reused within the TAMA.

Throughout the remainder of the analysis, $75 \times 10^6 \mathrm{m}^3 \cdot \mathrm{yr}^{-1}$ is used as the average rate of natural recharge (see above). The regional (TAMA) allotment of CAP water is $265 \times 10^6 \mathrm{m}^3 \cdot \mathrm{yr}^{-1}$, although a portion is not yet in use due to capacity constraints among local water providers. Tucson Water and other regional water purveyors are accelerating efforts to eliminate these physical limitations to CAP use. Although the regional CAP allotment is projected to expand to $323.8 \times 10^6 \mathrm{m}^3 \cdot \mathrm{yr}^{-1}$, competition among western states for Colorado River water suggests that increases much beyond that figure are unlikely. Only the quantity of wastewater reclaimed and reused seems

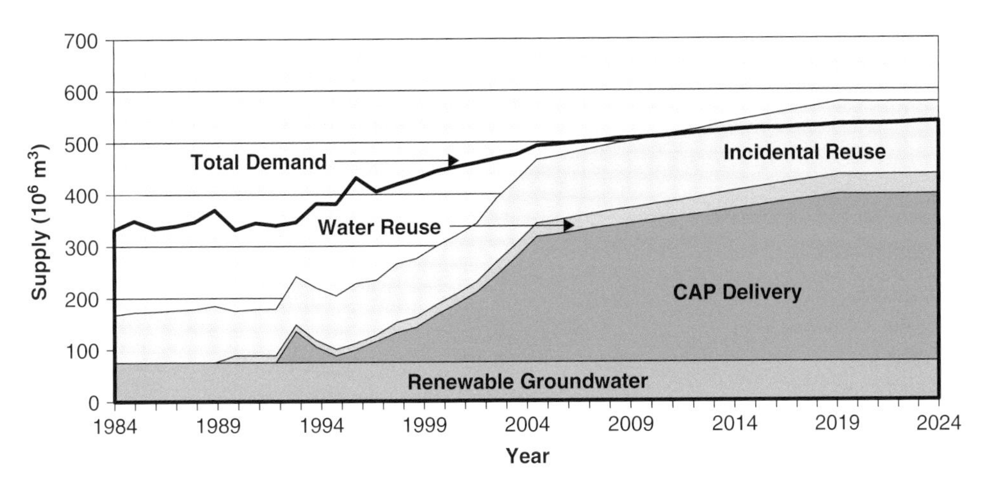

Figure 4 Water supply data and projections for the TAMA. The uppermost line represents the total rate of supply. Shaded regions correspond to the four major components of supply. Projected growth in the regional supply is primarily due to use of CAP water. The rate of groundwater mining is the difference between the demand and supply curves

elastic, or capable of expansion to meet safe-yield requirements. Source-specific water supply functions for the period after 1984 are provided in Figure 4.

In 1995, municipal water demand was $191.8 \times 10^6 \text{m}^3$ in the TAMA, and $85.1 \times 10^6 \text{m}^3$ of wastewater effluent was produced. Thus, 56% of the water produced for municipal use was consumed or otherwise lost through use, wastewater collection, treatment and disposal. A much greater percentage of water delivered to industry and agriculture is consumed. While these numbers might be refined, major implications for sustainable regional water supply do not depend on greater accuracy. It is estimated that from 1940 to 1997, the net loss of stored groundwater from the TAMA was 7.5–10 km^3. This represents 9–11% of the original volume of groundwater at depths less than 360 m below the land surface. The analysis to this point provides a rough check on that figure. That is, the volume of groundwater mined annually is the difference between the total demand and supply functions (Figures 3 and 4). The cumulative deficit from 1985 to 1997, calculated using the demand and supply figures provided, was about 2.5 km^3 (Figure 5). It is also evident from the figure that the annual water deficit was essentially steady from 1985 to 1997, and that efforts to introduce CAP water use are likely to reduce the annual water deficit dramatically. Most of the 7.5–10 km^3 groundwater deficit in the TAMA occurred prior to 1984, during a period of much greater agricultural demand in the TAMA and well before the initiation of Colorado River water delivery and water recycling programmes.

3.4 Sustainability considerations

Although the maximum amount of water that can be recovered and reused is less than half the municipal demand, reclaimed water that becomes a part of the regional

Table 6 The Year 1995 TAMA water budget (demand and supply). All figures are in millions of cubic metres per year. The visual (insert) illustrates relationships among supply/demand data leading to the estimate of mined groundwater

	Demand	*Supply*		
		Cap water	*Effluent*	*Groundwater*
Municipal Sector	191.8	0.12	9.5	182.2
Agricultural Sector	120.9	0	2.22	118.7
Industrial Sector	74.3	0	0.99	73.3
Loss due to Evaporation	4.56	0	0	4.56
TOTALS:	**392**	**0.12**	**12.7**	**378.7**
Groundwater Use				378.7
(Less) Net Natural Recharge				−75
(Less) Incidental Recharge[1]				−101.5
Groundwater Overdraft				**202.2**

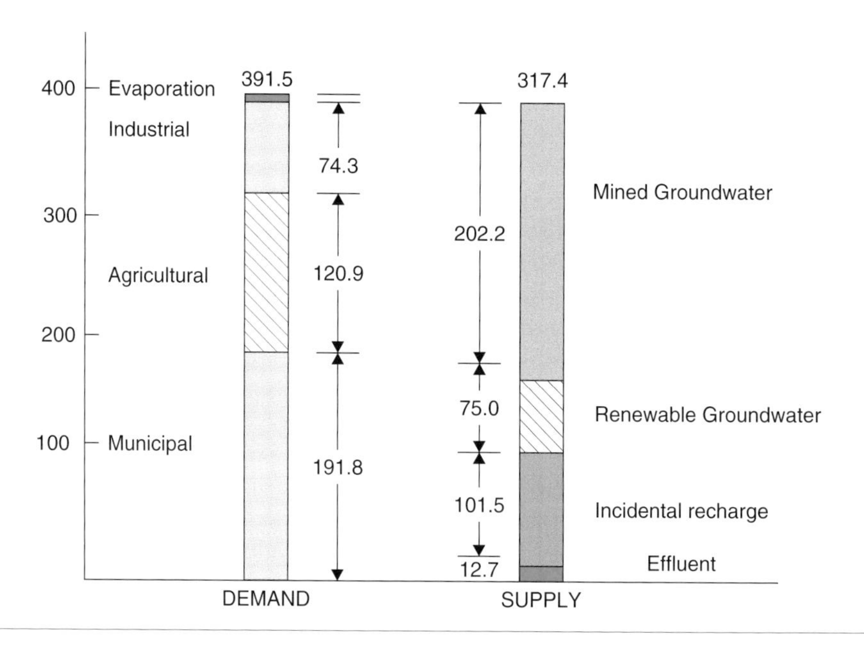

[1]Incidental recharge arises from fractional infiltration of water demands in municipal, agricultural and industrial sectors. Prescriptive assumptions are provided by ADWR. See below.

municipal water supply can itself be reused. To better determine the eventual limits to water supply in the TAMA, Figure 6 provides the basis for a detailed analysis of water use and safe-yield conditions. The ADWR has acknowledged that fractions of the water used in agriculture, industry and municipal water supply are returned to the regional aquifer as infiltrate. Estimates for agricultural (I_a), industrial (I_i) and municipal (I_m) fractions are 0.20, 0.12 and 0.04, respectively. Furthermore, an estimated 90% of the municipal wastewater effluent that is discharged to the Santa Cruz River infiltrates to

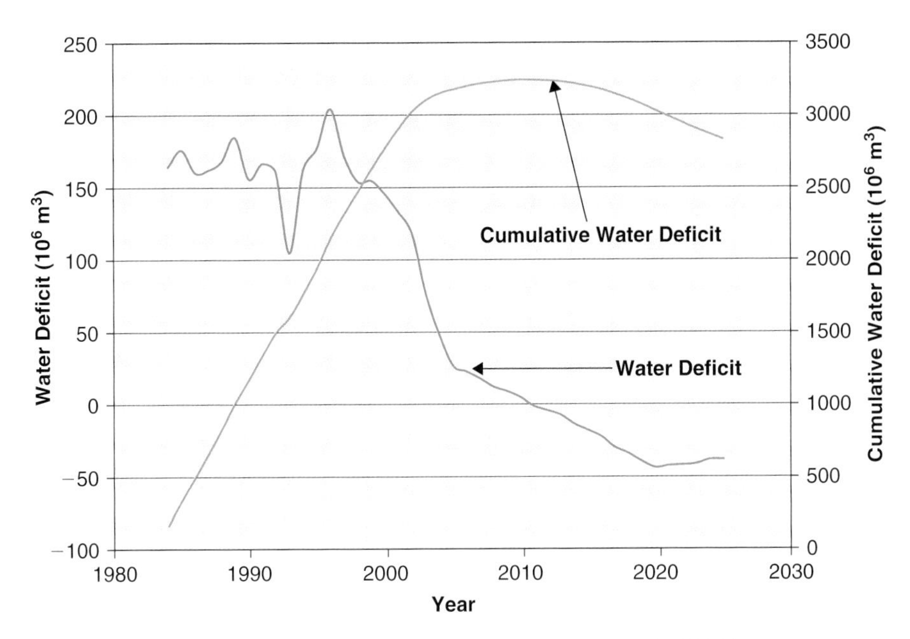

Figure 5 Annual and cumulative water deficits (groundwater mining) in the TAMA

groundwater within the TAMA boundaries ($I_e = 0.90$). Thus, a detailed water balance for the TAMA groundwater resource should have the following form:

$$A = S_{gs} + S_c + (RD_m - E) + I_a D_a + I_i D_i + I_m D_m + I_e E - D_a - D_i - D_m \quad (2)$$

where A is the net annual accumulation of groundwater in the TAMA
S_{gs} is natural rate of groundwater replenishment ($75 \times 10^6 \text{m}^3 \cdot \text{yr}^{-1}$)
S_c is the volume of CAP water supplied to the TAMA ($\leq 328.8 \times 10^6 \text{m}^3 \cdot \text{yr}^{-1}$)
R is the fraction of municipal water demand that is recovered, treated and reused or discharged as effluent (0.44)
D_m, D_i, and D_a are annual municipal, industrial and agricultural demands
I_m, I_i, and I_a are the fractions of municipal, industrial and agricultural demands that re-enter the groundwater as infiltrate (0.04, 0.12, 0.20)
E is the annual volume of wastewater effluent discharged to the Santa Cruz River and
I_e is the fraction of effluent that infiltrates to groundwater in the Santa Cruz River channel (0.90).

Overdrafts, or groundwater mining, would produce negative accumulation values.

Equation (2) also provides a means for recalculating the maximum sustainable rate of water use in the TAMA. The water balance itself is relatively easy to envision since there is very little direct surface water delivery in the TAMA. Under steady conditions, $A = 0$. Using the Figure 6 control volume, it is evident that a steady condition is achieved when

$$S_{gs} + S_c = (1 - I_i)D_i + (1 - I_a)D_a + (1 - I_m - R)D_m + (1 - I_e)E \quad (3)$$

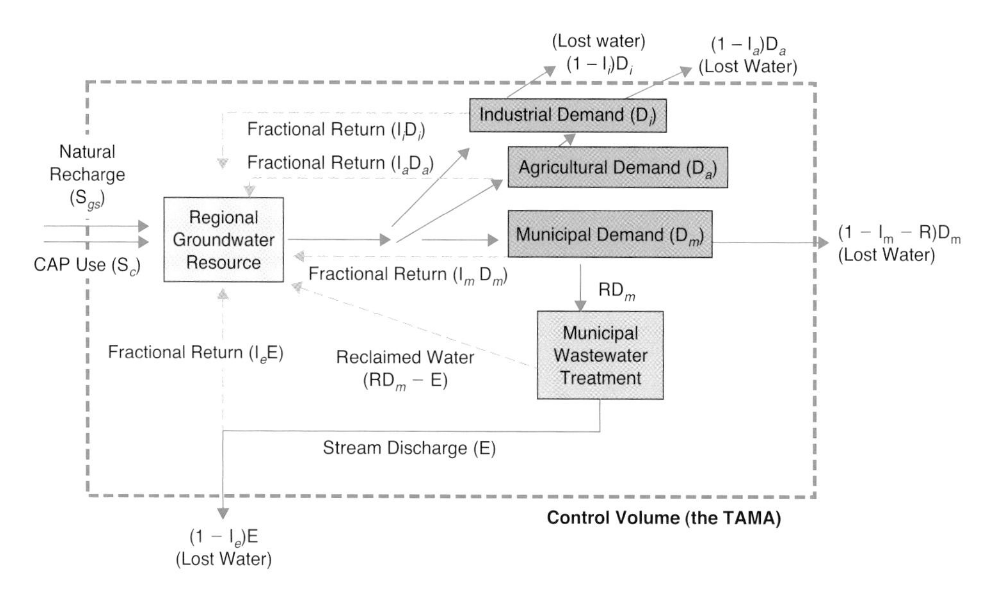

Figure 6 Relationship between water supply and wastewater treatment/reuse in the Tucson Active Management Area. For an explanation of symbols used, see the text. The dotted boundary represents the TAMA control volume used for water balances and sustainability calculations

Total water demand is $D_i + D_a + D_m$, so that under steady (sustainable) conditions:

$$D_i + D_a + D_m = S_{gs} + S_c + I_iD_i + I_iD_a + (I_m + R)D_m + (I_e - 1)E \tag{4}$$

To maximize the sustainable rate of water use, the TAMA should use its entire CAP entitlement ($S_c = S_{c,\max}$) and recycle all municipal wastewater effluent ($E = 0$). Then

$$(1 - I_m - R)D_m = S_{gs} + S_{c,\max} - (1 - I_i)D_i - (1 - I_a)D_a \tag{5}$$

$$\text{or } D_m = \frac{S_{gs} + S_{c,\max}}{1 - I_m - R} - \frac{(1 - I_i)D_i}{1 - I_m - R} \frac{(1 - I_a)}{1 - I_m - R} D_a \tag{6}$$

in which $R = 0.44$ and $S_{c,\max} = 328.8 \times 10^6\,m^3 \cdot yr^{-1}$, etc.

If our objective is to maximize the amount of water that can be served to the public, then clearly the community should try to limit industrial and agricultural demands. Since water recycling is possible following municipal use, this strategy would also maximize the total amount of water delivered. If industrial and agricultural uses in the TAMA were somehow eliminated, the maximum rate of water service would be about $765 \times 10^6\,m^3 \cdot yr^{-1}$. At a per capita rate of municipal demand equal to 651 LPCD, the largest population that could be sustainably served under this extreme condition would be about 3.2 million people. Realistically, agricultural and industrial demands will continue at or near current values. If industrial and agricultural demands are each assumed to continue at about $100 \times 10^6\,m^3 \cdot yr^{-1}$, then the maximum sustainable rate of municipal water service would be about $450 \times 10^6\,m^3 \cdot yr^{-1}$, and only about

1.85 million people could be served. The current population of the TAMA is estimated at 1.05 million.

3.5 Demand and supply projections

Historic water demands, wastewater discharges, and the State's estimates for fractional infiltration were used to project the regional demand and supply functions to 2025 (Figures 3 and 4). Similarly, groundwater overdrafts are projected in Figure 5. Water demand in the TAMA was projected for agricultural, industrial and municipal sectors using the following assumptions:

1. For agricultural water use:
   - The extent of irrigated lands will decrease by 50% from 1995 to 2025. This is consistent with an assumption that residential growth will continue to displace agriculture in the TAMA over the next 20 years.
   - The cropped acreage of agricultural land increased from 55% in 1995 to 65% in 2005 and has remained constant thereafter.
   - The average rate of consumptive water use in irrigated land will remain at $4.4 \times 10^3 m^3 \cdot yr^{-1}$per acre.
   - Demand for irrigation water on Indian land in the TAMA will reach $19.5 \times 10^6 m^3 \cdot yr^{-1}$ by 2010, and remain constant at that level until 2025.
2. For industrial water use:
   - Water use by the mining industry will be constant at $19.5 \times 10^6 m^3 \cdot yr^{-1}$ to 2025. This is considerably less than the current mining industry water right ($76.5 \times 10^6 m^3 \cdot yr^{-1}$), and water use in the mining industry is likely to fluctuate with the health of the global copper market.
   - Industrial turf-related water use will decline continuously until 2025, but other industrial uses will increase by a similar amount, so that total industrial demand will remain essentially constant.
3. For municipal water demand:
   - Per capita water demand is projected to 2025 at 651 LPCD, except for Native Americans on Indian lands, who will use 190 LPCD.
   - Municipal demand that is satisfied using reclaimed water is projected separately, then added to water demand, which was calculated as per capita demand times the projected population.

Population in the TAMA is predicted to grow to more than 1.4 million by 2025. Using a steady per capita demand of 651 LPCD, total municipal demand will reach about $330 \times 10^6 m^3$ in 2025 (Figure 2). It was also suggested that demand for municipal effluent, mostly for turf-related facilities, will reach almost $30 \times 10^6 m^3$ by 2025, making the total municipal water demand about $360 \times 10^6 m^3$. A complete discussion of assumptions employed in projecting regional water demand is available in the *Third Management Plan for Tucson AMA, 2000–2010* (ADWR, 1996).

The pace at which CAP water is introduced to the region will determine the eventual magnitude of the regional groundwater overdraft. Tucson Water has aggressively pursued the use of CAP water. The city's timetable for incorporation of CAP water into the regional water supply is probably the most realistic indication of the pace of CAP water utilization. The Tucson Water allocation of CAP water is about two-thirds

Table 7 Historic and projected effluent production rates of wastewater treatment facilities in the TAMA. All figures are in millions of cubic metres per year

Year	*All plants*	*Roger and Ina Road WWTPs (only)*
1990	72.9	68.8
1995	84.6	80.7
2000	95.1	90.7
2005	105.1	100.2
2010	115.2	109.5
2015	125.2	118.8
2020	135.2	127.9
2025	144.8	136.9

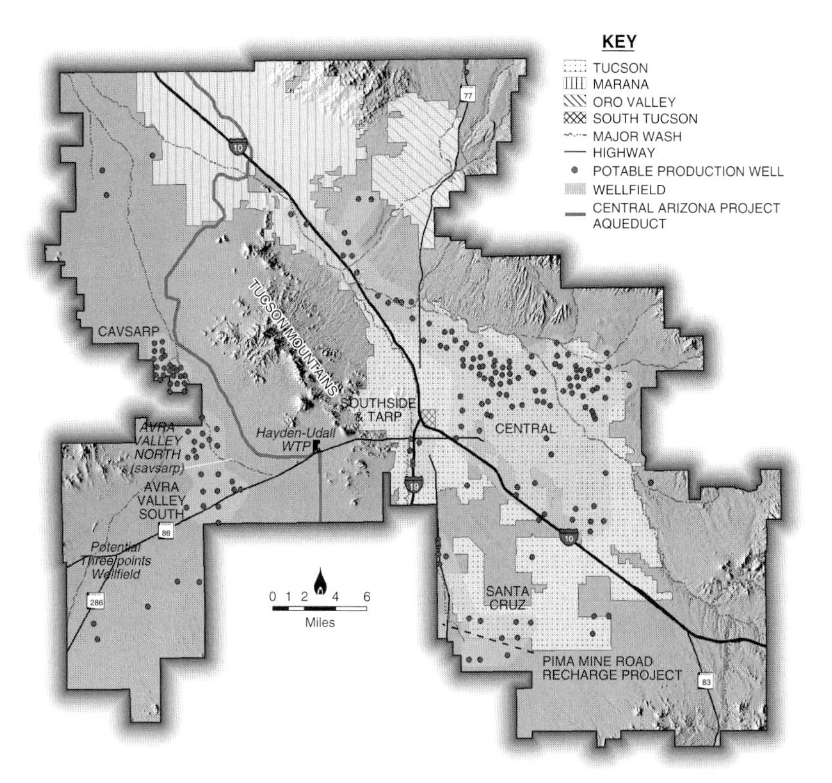

Figure 7 Major components of Tucson Water's water supply system. The map encompasses the region for long-range integrated water supply planning in the Tucson area (See also colour plate 20)

Source: Courtesy of Tucson Water

of the total allocation among all TAMA right holders, so that the utility's efforts will have a major effect on regional CAP use and groundwater overdraft.

Tucson Water favours recharge and recovery of CAP water before it is served to the public, although the Hayden-Udall Water Treatment Plant, the city's mothballed water

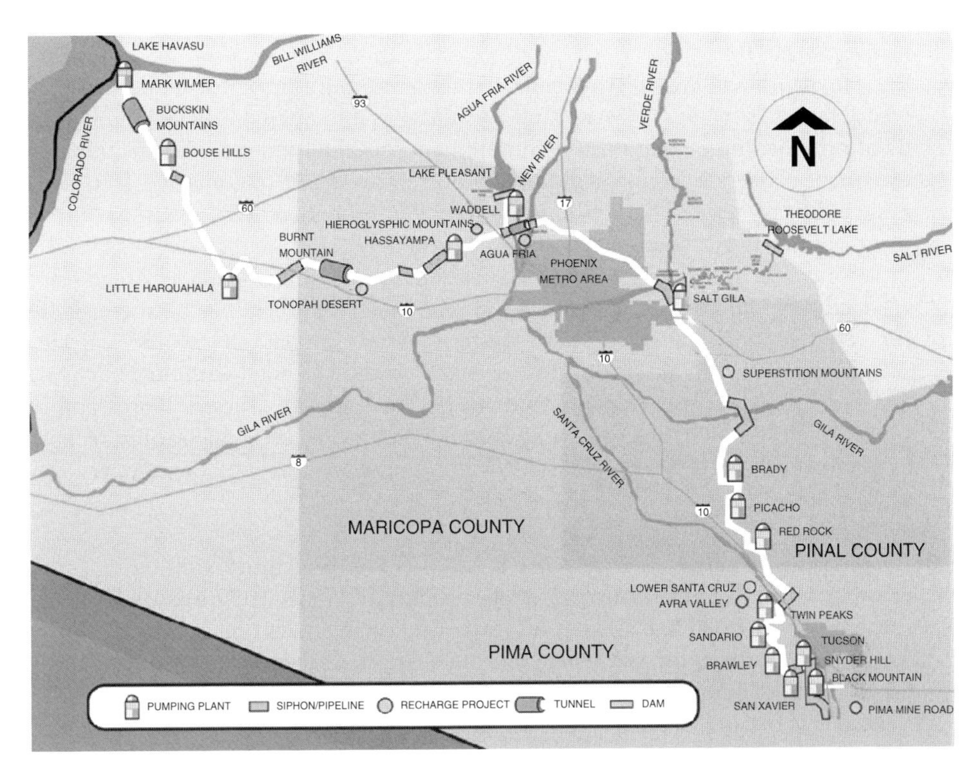

Figure 8 Layout and major features of the Central Arizona Project system. Colorado River water is mixed with water from the Agua Fria River in Lake Pleasant near Phoenix (See also colour plate 21)

Source: Courtesy of Central Arizona Project

treatment facility, could be revitalized for surface treatment and distribution of CAP water. That possibility is under examination along with other means to accelerate the use of CAP water. Major components of existing and proposed facilities for CAP underground storage and distribution, shown in Figure 8 are discussed below:

1. The Central Avra Valley Storage and Recovery Project (CAVSARP). This is a system of large-scale infiltration basins and recovery wells that were constructed to recharge and recover up to $75 \times 10^6 \mathrm{m}^3 \cdot \mathrm{yr}^{-1}$ of CAP water. The facility consists of $1.34\,\mathrm{km}^2$ of recharge basins, 27 recovery wells, a booster station, a 30 million litre reservoir and associated piping. Through CAVSARP operations, CAP water is blended with native groundwater prior to recovery and utilization. Regular deliveries of blended water began in 2001. The CAVSARP facility will soon be expanded to permit recharge and recovery of $100 \times 10^6 \mathrm{m}^3 \cdot \mathrm{yr}^{-1}$.
2. The Southern Avra Valley Storage and Recovery Project (SAVSARP). SAVSARP is planned for underground storage and recovery of $55–125 \times 10^6 \mathrm{m}^3 \cdot \mathrm{yr}^{-1}$ of CAP water. The project will include $0.8–1.6\,\mathrm{km}^2$ of infiltration basins, recovery wells, a booster station and pipelines for water collection and transport. Necessary

construction is likely to be in two stages. Current plans call for completion of this facility by 2010.

3. The Hayden-Udall Water Treatment Plant. Plant modification for the treatment of blended waters from CAVSARP and SAVSARP is anticipated. The available treatment options range from disinfection alone to desalination plus disinfection. The treatment process selection will depend on the results of ongoing work to determine public preference and willingness to pay. A more detailed discussion of treatment options is provided in the section devoted to water quality.
4. Other physical options for CAP water use. Additional underground storage/recovery facilities could be developed. The Three Points Well Field has been considered for construction in the south-western Avra Valley. It would produce $2.5–7.5 \times 10^6\,m^3 \cdot yr^{-1}$ of groundwater that would be blended with the CAVSARP/SAVSARP product waters prior to delivery. Furthermore, Tucson Water and the CAWCD jointly own the Pima Mine Road Recharge Project, which could recharge as much as $40 \times 10^6\,m^3 \cdot yr^{-1}$ of CAP water. There are currently no recovery wells at that site.

Full construction of just the CAVSARP and SAVSARP facilities will allow Tucson Water to store CAP water in excess of its (eventual) $179 \times 10^6\,m^3 \cdot yr^{-1}$ allocation. Thus, the city may be able to take additional CAP water when there is excess in the Colorado River system, for use when shortages are declared. A discussion of uncertainties in the CAP supply is provided elsewhere in the chapter.

In summary, Tucson Water has investigated a number of scenarios for the development of water supplies in its service area. All depend on using the city's CAP allotment as rapidly as possible to minimize groundwater overdraft. The construction of facilities necessary to bring this about can be completed by about 2010. Others in the TAMA will be slower to fully utilize their CAP entitlements. The difference between the projected demand (Figure 3) and supply functions (Figure 4) represents the rate at which water must be mined to satisfy water demand in the TAMA (Figure 5). The figure implies that groundwater mining will continue for just a few more years and at a much reduced rate. After about 2010, supply and demand will be in balance, at least temporarily. The total volume of groundwater mined over the period of the analysis (post 1984) will not exceed $3\,km^3$, if utilization of the regional CAP allocation proceeds at the pace currently envisioned.

The brief analysis also suggests that Tucson can continue to emphasize water conservation or minimize consumptive water uses to minimize groundwater mining. In addition, the public must be made aware that sustained development in the TAMA depends on treatment and reuse of wastewater. That is, treated wastewater will re-enter the regional water resource portfolio – whether through managed reuse or incidental groundwater recharge leading to eventual recovery and reuse. Only the acceptable uses for reclaimed water and the use-dependent degree of treatment remain to be determined.

Wastewater collection, treatment and discharge leading to eventual reuse would redistribute water within the TAMA. That is, water is used and collected predominantly in the Tucson Water service area but recharges the aquifer in the north-west quadrant of the TAMA (Figure 1). Because the resultant distribution of physical water does not match the projected distribution of municipal water demand in the TAMA, some form of regional water reclamation and managed reuse is likely in the future.

4 INSTITUTIONAL CONSTRAINTS – WET WATER VERSUS PAPER WATER

In their approach to safe yield, the State of Arizona makes a distinction between the physical or wet water supplies used in water balances to this point and paper water accounting rules. The AWS Rules actually govern paper water accounts, and application of those rules leads to a paper water balance. Several features of the AWS Rules are at odds with physical water balances, as is discussed at the end of this section. Legal/administrative restrictions to water distribution and use in the TAMA must serve social goals and protect water rights on a sub-regional scale, in addition to providing a rough regional hydrologic balance. Because they protect water rights, such 'paper' water constraints are by necessity more complex than the broad safe-yield requirements of the GMA. A detailed exposition of paper water restrictions in the TAMA is beyond the scope of this review. Therefore, binding rules and agreements that govern the distribution of water within the TAMA are briefly reviewed without attempting to provide an integrated perspective or describe the consequences of paper water rules to individual water purveyors.

4.1 Federal water constraints

The Colorado River Compact of 1922 allocated waters of the Colorado River among seven states, including four of the so-called Upper Basin states (Wyoming, Colorado, Utah and New Mexico), the Lower Basin and Mexico. The Lower Basin states are California, Arizona and Nevada. To make the agreement work, the Upper Basin is required, to release 9.25 km^3 of Colorado River water past the gauging station at Lee's Ferry, the gauging station just upstream from Lake Mead, for use by the Lower Basin states, unless a shortage is declared on the Colorado River by the Department of Interior. Of the 9.25 km^3 for the lower states, California is entitled to 5.4 km^3, Arizona to 3.5 km^3 and Nevada to 0.4 km^3. Ratification of the Compact by 6 of the 7 states triggered construction of the Hoover Dam and the All American Canal, which transports water into southern California. Arizona did not sign the agreement until 1944. Unfortunately, shortage provisions may be more critical than originally envisioned due to possible overestimation of Colorado River flows or even further reduction of such flows due to climate change in the western United States.

Because the Colorado River was remote from its major population centres, Arizona at first used just a fraction of its water right. The Colorado River Basin Project (CAP) Act of 1968 resulted in the construction of the Central Arizona Project, which brought Colorado River water across the state and past Phoenix and Tucson. Of the 3.5 km^3 of Colorado River water allotted to Arizona annually, 1.9 km^3 is reserved for the CAP, including $266 \times 10^6\,m^3 \cdot yr^{-1}$ within the TAMA, allowing Arizona to approach full utilization of its rights to Colorado River water.

The Tohono O'odham Nation and Pasqua Yaqui tribe are located in the TAMA. A negotiated settlement between the Tohono O'odham Nation and major water users of the TAMA (the Southern Arizona Water Rights Settlement Act of 1982, SAWRSA) obligated the Secretary of Interior to provide $34.8 \times 10^6\,m^3 \cdot yr^{-1}$ of TAMA effluent to the Nation. This must be provided by Tucson Water. More than 20 years later, the lawsuit that motivated this agreement remains unsettled. Dismissal awaits an act of Congress that would place the settlement into a broader context involving other

Native American water rights. In addition, the Tohono O'odham Nation holds rights to $46.6 \times 10^6 m^3 \cdot yr^{-1}$ of CAP water.

4.2 State water law

Major provisions of the Arizona GMA have already been covered. The GMA and pursuant legislation established the legal and administrative frameworks for management of water service to 80% of the State's population. Briefly, the primary management goal for each of Tucson's AMAs is a hydrological balance between demand and supply, or attainment of safe-yield requirements, by 2025. The most important tools for the pursuit of water balances are the AWS Rules, which prohibit development that fails to show how water demands can be met for the next 100 years using renewable water supplies. Only utilities that themselves meet State AWS requirements can extend water service within their respective service areas. Many of the practical difficulties arising from AWS demonstrations – and particularly those arising from lack of direct access to Colorado River water – can be circumvented through contractual arrangements with the CAGRD, which provides paper groundwater rights by obtaining and recharging unused Colorado River water. Local hydrological imbalances frequently attend such recharge activities. Nevertheless, the CAGRD plays an important role in the State, mitigating the tendency of AWS Rules to impede desirable economic growth. Allowable groundwater (mining) provisions of the AWS Rules (see above) also provide flexibility by establishing the volume of water that can be mined before AWS constraints become binding.

The CAGRD has also attempted to identify water rights that are potentially available for purchase or lease within the State. This tends to facilitate assignment of the State's limited water resources to the highest economic or social uses. Primary sources of underutilized water rights include waters reserved for use by Native Americans and agriculture along the Colorado River (non-CAP water). The CAGRD suggested that reallocation of 20% of Arizona's Native American rights to Colorado River water and 20% of the water used along the Colorado River in Arizona would liberate $>0.5 km^3 \cdot yr^{-1}$ for other uses.

4.3 Local agreements

The most important of the local inter-government agreements (IGA) affecting water supply in the TAMA was reached by the City of Tucson and Pima County in 1979. The IGA transferred city-owned sewage to Pima County, and the city received rights to 90% of residual effluents from metropolitan treatment plants, after the Department of the Interior (Tohono O'odham Nation) received the $34.8 \times 10^6 m^3 \cdot yr^{-1}$ required under SAWRSA. The county retained rights to the remaining 10% of the effluents from these facilities. The agreement cleared the way for Tucson Water to market reclaimed water for landscape irrigation and consolidated water resources planning within the city. Additional agreements among the City of Tucson and other municipal water purveyors in the TAMA divided effluent rights based on potable water deliveries.

When Tucson Water initiated local delivery of CAP water in 1992, centralized treatment was provided at the Hayden-Udall Water Treatment Plant, and treated CAP water was distributed over much of the service area using the existing water distribution system. The sudden influx of Colorado River water produced high velocities and even local flow reversals that scoured precipitates and/or corrosion products from existing pipes. As a consequence, Tucson Water almost immediately stopped taking CAP water and temporarily returned to full reliance on groundwater. A Tucson citizen initiative, the Water Consumer Protection Act of 1995, was passed that prohibited direct delivery of Colorado River water for five years unless it was treated to achieve a quality equal to that of local groundwater. Since this was not economically feasible, the initiative was effectively a prohibition against direct delivery of CAP water. A proposed extension of the initiative was defeated in 1999, and most provisions of the Act expired leading to the city's current effort to accelerate the rate at which CAP water is utilized.

4.4 Differences between wet and paper water balances

The most obvious departure of the AWS Rules from the wet water balance developed earlier is the State's refusal to acknowledge formally that regional aquifers experience a natural, predictable rate of replenishment ($75 \times 10^6 \, m^3 \cdot yr^{-1}$ in the TAMA). Natural replenishment will add perhaps $7.4 \, km^3 \cdot yr^{-1}$ to the TAMA groundwater resource during the AWS planning period of 100 years. Furthermore, State accounting provides only a 50% credit for effluent discharged from wastewater treatment plants along the Santa Cruz River, although 90% recharge is probably more representative. Again, this leads to significant practical differences in wet and paper water accounts.

To offset these differences and to provide flexibility in meeting AWS requirements, the State allows utilities to include a specified volume of groundwater in their paper water accounts. This is known as 'allowable mined groundwater'. The volume of groundwater that can be mined while converting to renewable resources is equal to 15 times the year 1995 demand in the water providers' respective service areas. An additional volume of groundwater can be mined when providers purchase and extinguish grandfathered groundwater rights – agricultural or industrial water rights that are not subject to AWS provisions or safe-yield determinations required by the GMA.

In the TAMA, groundwater mining accounts became active for Tucson Water in 2001, and for five other water providers in 1999. Mined groundwater allowances for these six utilities total $2.3 \, km^3$. Tucson Water can take an additional $2.5 \, km^3$ without replacement because the department purchased Avra Valley land (Figure 1) and extinguished associated agricultural (grandfathered) water rights. Finally, there is no requirement to replace groundwater pumped for the treatment of hazardous waste remediation projects. This provision of the AWS Rules is expected to result in less than $0.25 \, km^3$ of groundwater use in the TAMA before 2025. Thus, the total volume of allowable mined groundwater in the TAMA is about $4.9 \, km^3$, of which Tucson Water holds the major right. As a practical matter, the allowable mined groundwater provisions of the AWS Rules tend to offset the State's unwillingness to include natural groundwater replenishment among paper water supplies, so that paper water balances may provide a long-term hydrological balance in the TAMA.

5 WATER QUALITY CONSIDERATIONS

5.1 Salt

Delivered water quality will change as the region shifts from exclusive reliance on groundwater to full utilization of its CAP water right. The situation is best illustrated by analyzing the implications of the shift to CAP water on the quality of water delivered by Tucson Water.

The most striking difference between CAP water quality and that of the principal groundwater sources in the TAMA is in mineral content (Table 1). The average total dissolved solids (TDS) concentration in groundwater delivered by Tucson Water in 2000 was ≤300 mg/L. In 1995, the average salinity in the Colorado River below Parker Dam, the point at which water enters the CAP canal, was 775 mg/L. As demands on the Colorado River increase, its salinity is predicted to increase. Even the current TDS level in the Colorado at the CAP intake depends on a degree of federal and state salinity management, without which the expected TDS concentration would exceed 900 mg/L. En route to Tucson, Colorado River water is mixed to a degree with low-TDS water from the Agua Fria River in Lake Pleasant to yield a blend with a slightly lower TDS at the Tucson turnout of the CAP canal. During 2000–2005, TDS concentrations in CAP water at the Tucson turnout ranged from 550 to 680 (Figure 9). During that period, the general trend in salt content was upward.

Although trends in Colorado River water mineral content are important, they do not affect the most general conclusion – without salt management steps, the replacement of groundwater with CAP water from 2000 to 2020 will gradually increase the TDS of water served in the TAMA. The method of utilizing CAP water adopted by Tucson Water was designed to avoid an abrupt TDS increase. Because the intakes of CAVSARP recovery wells are well below the local water table, recovered water is a blend of infiltrated CAP water and natural groundwater. However, with an ever increasing CAP volume contribution, TDS concentrations in the blended supply, known as Clearwater, are increasing and will ultimately match CAP concentrations. The salinity of recovered water in several CAVSARP wells already approaches that of CAP water (Figure 9).

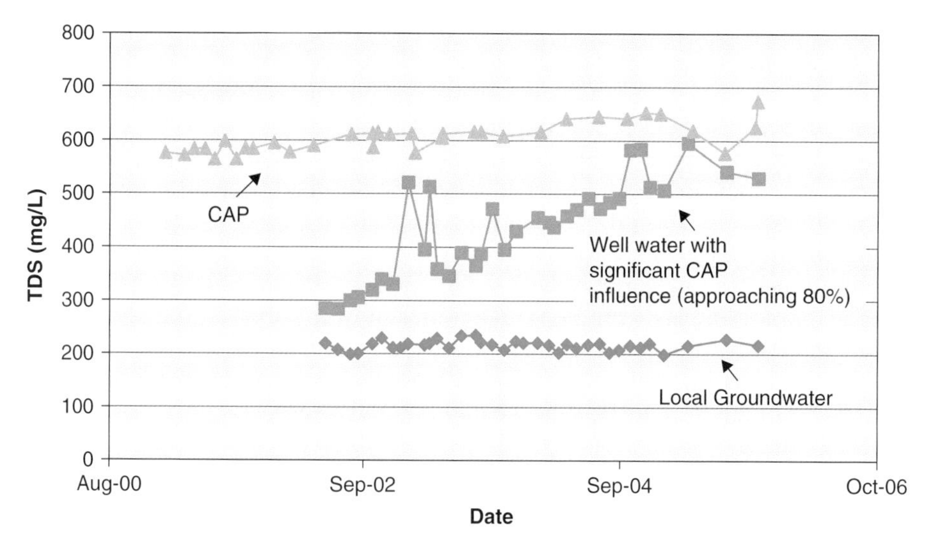

Figure 9 Record of TDS levels in CAP water (southern, Tucson terminus of the CAP canal), local groundwater in the Avra Valley and water in a CAVSARP recovery well. Increasing salinity in the well water reflects the increasing percentage of CAP water in the recovered blend

Current treatment of Clearwater consists of pH adjustment and chlorination before it enters the Tucson Water distribution system. Tucson Water is currently evaluating salt removal options prior to water delivery, due to evidence of customer preference for lower salinity water. The city's primary alternatives are:

- Salt removal from at least a fraction of the recovered CAVSARP water and blending to maintain TDS levels $\leq$450 mg/L.
- Continued infiltration and recovery without treatment – essentially allowing salinity to find its own level in the blend of water sources used to satisfy demand.

In the latter case (no treatment), it is possible to project an average TDS in the regional groundwater supply, considering the likely time-dependent contributions of CAP water, native groundwater and reclaimed water to the groundwater resource. Rather than expose an entirely new set of assumptions/projections that pertain to waters delivered within the Tucson Water service area, however, a salt balance is developed for the TAMA water supply as a whole. It is emphasized that projections represent regional averages and that TDS concentrations at delivery points may be significantly lower or higher. The following tributary projections and assumptions were made:

- The TDS of water reaching the Tucson terminus of the CAP canal is constant at 650 mg/L.
- Utilization of CAP water in the TAMA follows the schedule provided in Figure 4.
- Water use and collection/reclamation adds 250 mg/L to the TDS concentration in recovered water.
- Mined groundwater required to produce a water balance in any given year has a TDS concentration equal to that of the regional groundwater average (i.e. it increases over time).

To project the possible magnitude of TDS changes in the regional groundwater resource, it is assumed that the entire mineral content of water utilized regionally is retained in the TAMA. In a worst case scenario, all of this salt is transferred to the regional groundwater supply. That is, there is no major efflux of salt from the TAMA nor any appreciable accumulation of salt in vadose zone soils. The mass of salt added to the aquifer system each year would be equal to the volume of total water demand times the average TDS concentration of the water served. In 2025, for example, this would amount to 3.13×10^5 MT, or 857 MT per day, of which roughly 55% would come from the use of CAP water.

It is also necessary to select a mixing volume for the regional groundwater basin. Here a conservatively large volume was chosen, equal to the water volume in the top 300 feet of groundwater in the TAMA. The crudely estimated mixing volume is thus 20 MAF, or one-third of the TAMA groundwater volume between 90 and 360 metres below the land surface. Notice that this volume will decrease somewhat with time, as groundwater is temporarily mined to produce a regional water balance. Under those circumstances, a mass balance on salt yields:

$$V(\Delta \mathrm{TDS}) + \mathrm{TDS}\,(\Delta V) = S_C\,(650) + V_R\,(\mathrm{TDS} + 250) + S_{GS}\,(250) - D_T\,(\mathrm{TDS}) \qquad (7)$$

where V is the time-dependent volume of water in the well mixed zone (L)
TDS is the time-dependent salinity, (mg/L)
S_C is the annual volume of CAP water used, $(\mathrm{L \cdot yr^{-1}})$
V_R is the volume of water reused each year (including incidental recharge), (L)
S_{GS} is the natural recharge rate, $(\mathrm{L \cdot yr^{-1}})$ and
D_T is the total water demand, (L).

If all CAP water and reclaimed water is recharged prior to recovery and use, then $\Delta V = S_C + S_{GS} + V_R - D_T$. There was 24.7 km^3 of groundwater in the central volume at the start of the analysis (1993), and the initial average TDS was 280 mg/L. The procedure was used to project aquifer TDS concentrations from 1993 to 2010 (Figure 10).

The exercise suggests, not surprisingly, that salt levels will increase most rapidly after relatively large volumes of CAP water are utilized in the TAMA. The overall average TDS increase over the 18-year analysis is predicted to be almost 60 mg/L. More importantly, the increase in average groundwater salinity was predicted to increase by 5 mg/L·yr from 2005 to 2010. This is a consequence of near-full utilization of the TAMA CAP allotment and consequent salt accumulation. Without additional salt-management steps, the regional groundwater salinity will continue to increase at that rate, even after the groundwater TDS concentration reaches that of CAP water. Because this will take decades, salt-management decisions need not be immediate. Nevertheless, the analysis suggests that inevitable reliance on Colorado River water and reclaimed wastewater in the TAMA, without additional steps to remove salt, will produce a slow but continuous decline in regional groundwater quality. Local effects could be more rapid.

The alternative to allowing groundwater salinity to seek its own level, or salt management to maintain groundwater salinity at 450 mg/L, has its own difficulties. Facilities construction and operational costs for salt management over the next two

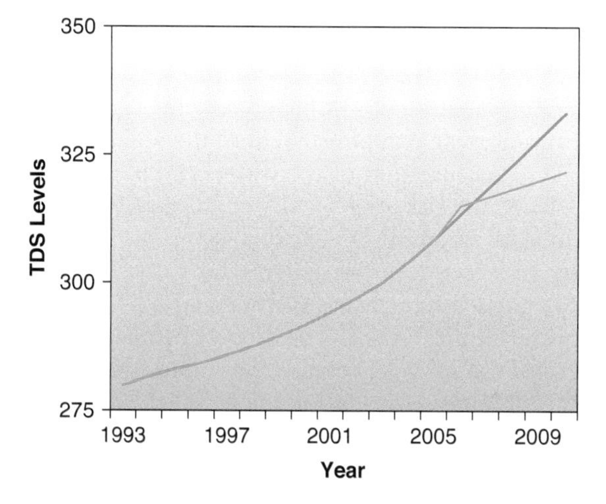

Figure 10 Projected TDS levels in TAMA groundwaters. Figures represent average concentrations in a control volume consisting of groundwater (initially) within the top 90 m of the regional aquifer system

decades have an estimated present worth of about a third of a billion dollars. Technologies for salt management have essentially no track record at the scale envisioned, and each carries with it new responsibilities for minimizing water loss as brine and stabilizing/disposing of brine salts. All of this is beyond the scope of the review, but illustrates the magnitude of salt-related technical issues that are being faced for the first time in the American south-west.

It is also possible to build a regional salt-management strategy based on reverse osmosis treatment of reclaimed water. Potable water reuse strategies would almost certainly include a reverse osmosis treatment step that would also lower the salt content from wastewater levels to 30–40 mg/L. Within the framework of the salt balance/TDS projection provided to this point, it is easy to evaluate salt management based on reverse osmosis treatment of wastewater that is intended for reuse, including incidental recharge. After modification to reflect this scenario, equation (7) becomes:

$$V\,(\Delta \mathrm{TDS}) + \mathrm{TDS}\,(\Delta V) = S_C\,(650) + V_R\,(40) + S_{GS}\,(250) - D_T\,(\mathrm{TDS}) \qquad (8)$$

Equation (8) was used to project the average time-dependent regional TDS level in TAMA groundwater (all other conditions, per above). Results are plotted in Figure 10. Reverse osmosis treatment of reclaimed water is assumed to begin in 2006 (it did not). The primary differences between the two simulations is that, with salt removal from wastewater prior to reuse, the average groundwater TDS will increase by only 1.7 mg/L·yr, or one-third the rate of TDS increase that is predicted in the absence of salt management. The same caveats that applied to the previous analysis apply here.

6 UNCERTAINTIES IN THE WATER MANAGEMENT PROCESS

The Tucson municipal area has reached a transition point with respect to development and use of water resources. The city and surrounding water providers are in the midst of research and analysis that will lead to full utilization of the regional CAP entitlement within the next half-dozen years. No one suggests that there is another rational course. There is also a growing awareness, however, that salt management is a related issue. CAP water will bring with it 180,000 MT of salt each year. Tucson residents must decide whether it is necessary and/or economically feasible to separate and dispose of salt to satisfy long-term salinity objectives. Treatment processes necessary for salt management are not widely practised on the scale envisioned for Tucson, and the region cannot afford to discard brines that contain more than a few percent of its CAP allotment.

The value of water in the Tucson area is an important component of planning for salt management. The city pays >\$80/1,000 m^3 to purchase CAP water, but would purchase more than its current allotment, if more water was available at that price. The value of water is certainly greater than \$80/1,000 m^3 in the Tucson area and elsewhere in the south-west. At the other extreme, high-demand residential customers pay >\$2,400/1,000 m^3 for the last increment of water taken due to the City's rate schedule, which is designed to discourage high-volume use. Since this figure includes the costs of treatment and distribution, it probably overestimates the marginal value of water to the region. If $800/100 m^3 is arbitrarily selected for the analysis, and just 10% of the regional CAP allotment is lost as brine, the contribution of water loss to the overall brine disposal cost would be about $26M per year. Clearly the decision to desalinate and the selection of desalination processes will have important economic repercussions for the community.

It is also clear that safe yield cannot be achieved in the TAMA without reclamation and reuse of municipal wastewater. Arguably, this is already occurring through in-stream incidental recharge. However, metropolitan Tucson must soon make decisions related to the nature of acceptable uses for reclaimed water that is formally added to the regional portfolio of water supplies and use-dependent standards for water reclamation and reclaimed water. Such uses and standards have been discussed in water-short areas of the United States for some time, yet no general guidance or set of principles has evolved, and western communities have come to very different conclusions.

In the last decade, the discussion of water reuse has focused, to a degree, on trace organics that are added to wastewater as a consequence of human use. Many of these survive conventional wastewater treatment, at least in part. Endocrine disrupting compounds (primarily oestrogens) are chief among the public's concerns. It has been widely shown that oestrogenic compounds in wastewater effluent can alter sexual development and behaviour in continuously exposed animals. The evolving body of scientific information concerning fate and the effects of trace anthropogenic compounds in wastewater is likely to colour debate over the propriety of water reclamation/reuse when that discussion occurs in Tucson. Until that debate occurs, however, the manner in which wastewater effluent will be incorporated into the regional water resource portfolio is uncertain.

Finally, there is uncertainty relative to the sustainability or true average yield of the Colorado River itself. When the Colorado River flow was divided among the seven states that eventually signed the Colorado River Compact, the best available hydrology indicated that the average flow in the river was about 20.4 km^3. A longer record, developed more recently from tree-ring data and other sources, suggests that there is considerably less water in the Colorado, perhaps as little as 16.7 km^3 on average. If so, water stored in the Colorado River impoundments will gradually diminish as states use a higher percentage of their respective entitlements under the existing agreement (9.3 km^3 to the upper states, 9.3 km^3 to the four lower states and 1.9 km^3 reserved for Mexico). Since there are currently about 32 km^3 of water stored in Lakes Mead and Powell, the process could take some time and may be difficult to observe on a time scale of a few years. Making the long-term availability of Colorado River water even more perilous are the possible effects of global warming on precipitation and evaporative losses in the Colorado River basin. Estimates of future average river flow as low as 6.2 km^3 have been offered based on greenhouse-driven climate models.

Rules governing the declaration of water shortages on the Colorado are under discussion among the parties to the Colorado River Compact. Such rules, which will probably be based on water levels in Lakes Powell and Mead, will trigger steps to conserve water within the river system – primarily reduced allocations to the Compact states. Final rules are likely within the next year. Any outcome that reallocates water during declared shortages, however, is likely to reduce the total CAP allocation below 1.9 km^3 due to its junior status among right holders. Although agricultural users of CAP water are junior to municipal water providers, a long-term or endemic shortage of stored water in the river system is almost certain to affect the entitlements of municipal water providers in the TAMA. Unfortunately, there is not yet a reasonable way to anticipate the magnitude or length of cutbacks derived from regional drought, although water providers should build contingency plans based on pessimistic scenarios.

7 CONCLUSION

Tucson and other high-growth metropolitan areas of the American south-west are in the process of deciding their long-term water futures. In some of these communities, continued growth is inevitable, so that waters of increasingly questionable initial quality will be added to their respective water resource portfolios. Where demand is projected to exceed supply, agricultural users will be persuaded to relinquish water rights to satisfy higher-valued needs, additional conservation measures will be adopted, or more creative solutions will be devised. Regulatory institutions must be sufficiently fluid to facilitate the transfer of water rights, especially in light of major resource uncertainties derived from the potential loss of flow in the Colorado River. There is sufficient water to satisfy municipal demand without sustained groundwater mining if cognizant institutions can facilitate the exchange of water rights.

Water quality presents a different picture. The lower Colorado River provides water with much higher salinity than that of Tucson's historic groundwater resource. Full utilization of rights held by TAMA water providers will bring about 180,000 MT of salt to the region each year. Unless steps are taken to manage salinity and find the means for permanent salt retirement, essentially none will leave the region. The predicted rate of salt accumulation in groundwater and soils suggests, however, that engineered solutions for salt-management can be implemented over a period of years to decades without unreasonable environmental deterioration. The chief difficulty in making salt-related decisions lies in assigning an appropriate economic value to water quality. The cost of specific salt-management measures is probably known, for example, within a factor of two. The present value cost of maintaining TDS levels at or below 450 mg/L in the Tucson Water service area was roughly estimated at a third of a billion dollars. While there are uncertainties in that figure and even in the technologies selected for salt management, estimation of benefits depends on the value assigned by the public to the maintenance of salinity levels. In the short term at least, the value of salt separation and disposal may be largely aesthetic, and the public's willingness to pay for aesthetic benefits, even in an area as central to quality of life as the water supply, is essentially unknown.

The questions presented by salt management today will be raised in a much more visceral way when applications for reclaimed water and related treatment or water quality requirements are selected. Water resource sustainability in the TAMA depends on water reuse. This is hardly an issue, and, although it is seldom mentioned, incidental groundwater recharge with treated wastewater is already a significant component of the regional water balance. There are complicating factors here, not the least of which is our growing awareness of trace organic contaminants that are not completely removed through conventional wastewater treatment. The extent to which these contaminants affect human and environmental health through wastewater-related exposure will be a source of discussion for decades and may never be completely resolved. It has been argued forcefully of late that the developed world tends to overspend in the environmental area to achieve health benefits that are largely illusory. That is, in purchasing higher levels of water treatment or restricting the acceptable uses of reclaimed water, we may be paying for peace of mind as opposed to material health improvements. If the public wants to purchase peace of mind, that intention will be honoured. Our need, then, is for an informed public that recognizes the importance of high quality water, appreciates that there are trade-offs in allocating money for public benefits, and makes deliberate decisions of modest technical uncertainty.

REFERENCES

In preparing this document, the authors drew heavily on the following documents.

Arizona Department of Water Resources. 1999. *Third Management Plan for Tucson Active Management Area*, 2000–2010.

City of Tucson Water Department. 2004. *Water Plan: 2000–2050*, Final Draft.

Gelt, J., Henderson, J., Seasholes, K., Tellman, B., Woodard, G., Carpenter, K., Hudson, C. and Sherif, S. 1999. *Water in the Tucson Area: Seeking Sustainability*. Water Resources Research Center, College of Agriculture, The University of Arizona.

Case Study III

Upper Awash River System in Ethiopia

Messele Z. Ejeta[1], Getu F. Biftu[2] and Dagnachew A. Fanta[1]

[1]*California Department of Water Resources, Sacramento, California, USA*
[2]*Golder Associates Ltd., Alberta, Canada*

1 INTRODUCTION

The Awash River Basin of Ethiopia lies between the Blue Nile basin to the north, the Ghilghal-Omo basin to the west, and the Rift Valley Lakes basin to the south, and drains in the Afar depression to the north-east. The river originates in the highlands of central Ethiopia at an altitude of about 3,000 metres above mean sea level. It drains the Batcho plains and flows south-west after which it enters the Great Rift Valley and then follows the valley and drains in Lake Abbé near the Republic of Djibouti, at an altitude of about 250 metres above mean sea level. The total length of the river is about 1,200 km and its catchment area is about 113,700 km^2 (FAO, 1964).

The Awash River course is divided into three major systems: Upper Awash, Middle Awash and Lower Awash. It covers a wide climatic zone, ranging from humid subtropical to arid zones. The basis for the classification of the Awash River Basin is mainly the water development activities in the three regions of the river course.

The major projects in the Upper Awash System consist of the Koka Dam and power generation, the Awash II and Awash III power generation stations, as well as the Wanji and Matahara sugarcane plantations and factories. The Wanji sugarcane plantation is currently 6,000 hectares and is expected to add a further 4,000 hectares of irrigated land. The Matahara sugarcane plantation is currently 12,000 hectares and is expected to add a further 5,000 hectares of irrigated land.

The major projects in the Middle Awash System include the Amibara Irrigation Project for cotton and banana plantations, as well as the new Kassam Dam and Sugarcane Project. The cotton and banana plantations occupy about 10,300 hectares of irrigated land with nearly three to one land use proportion, respectively. The Kassam Dam sugarcane plantation, which is being built, is expected to occupy about 30,000 hectares of irrigated land. A brand new sugar-producing factory is also expected to go operational soon in this location.

The major project in the Lower Awash System is the Tandaho Irrigation Project. A new water storage dam is also being built in this location to irrigate an expected 50,000 hectares of sugarcane plantation. An existing cotton plantation of about 10,000 hectares is planned to be converted to sugarcane plantation. A new sugar-producing factory is also being constructed in this location.

The Awash River Basin is unique in the fact that the river emanates and ends within the boundary of Ethiopia. This feature of the basin makes integrated management of its resources simpler than other trans-boundary rivers. Several towns and cities, including the capital, Addis Ababa, and industrial enterprises lie within the Awash River Basin (Hailemariam, 1999). Figure 1 shows a satellite land cover image of Ethiopia and Figure 2 show that of the Awash basin.

The Upper Awash System is one of the relatively well-developed river reaches in Ethiopia. The cities of Adama, Wanji and Matahara are located near or along the

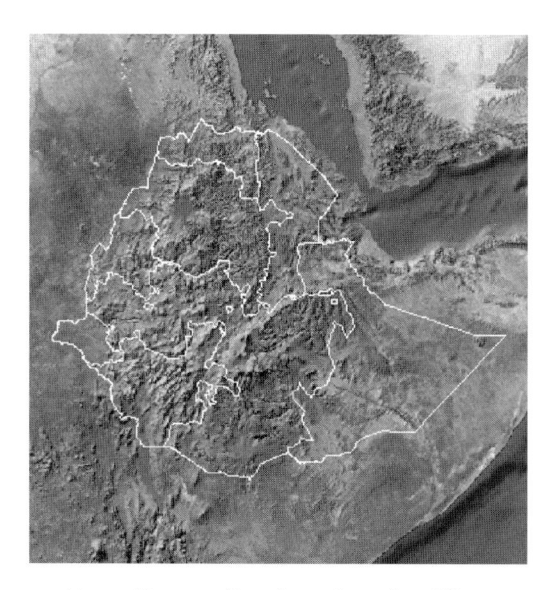

Figure 1 Ethiopia's topographic satellite map (See also colour plate 22)

Source: FAO

Figure 2 Landsat 7 Mosaic Cover of Awash Basin (See also colour plate 23)

Source: NASA

Awash River in the Upper Awash reach. The Koka Reservoir, which came online in 1961 with a net available capacity of 1,660 km^3 behind a concrete dam that is 42 metres high, is one of the major water supply facilities in the system. The normal annual outflow from the reservoir is about 120,000 cubic metres per second, whereas losses by evaporation and percolation are 31,500 and 38,000 cubic metres per year, respectively (FAO, 1964).

This case study attempts to assess the state of the water resources of the Awash River System and gives some perspectives for its integrated management. While considering the projects in the whole basin for integrated water resources management, this case study focuses on those projects in the Upper Awash System.

2 CLIMATE CHARACTERISTICS

Ethiopia is a country of great geographical diversity, with high and rugged mountains, flat-topped plateaus, deep gorges, river valleys and plains. The country has three major climatic regions as shown in Figure 3. The Upper Awash is located in Region A of Figure 3.

The climate of the Upper Awash region is dominated by its near equatorial location under the influence of the inter-tropical convergence zone and high altitude (ELC and Tropics, 2005). Long dry spells, high intensity and short duration rainfall, and occasional floods are some of the factors for consideration in the planning and management of the water resources of this region. In effect, the temporal variability in the climatic characteristics of the region is an important consideration for efficient reservoir operating rules for urban and agricultural water allocation, as well as for long-term water resources planning, design and construction of new water supply infrastructures. Based on the uncertainty of water availability, this region may be characterized as a semi-arid to arid region.

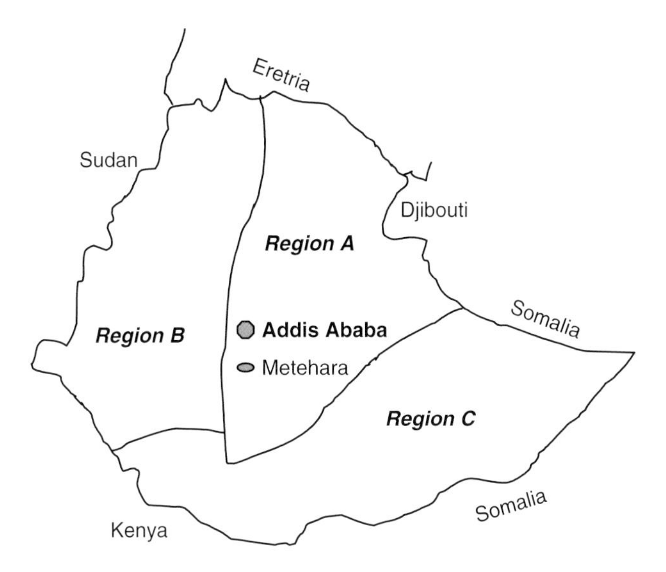

Figure 3 Major climatic regions of Ethiopia

Source: adopted from Bekele, 1993

Table 1 Seasonal variation of precipitation in Ethiopia

Climatic region	*Season*		
	Long rain	*Short rain*	*Dry*
Region A: central and eastern Ethiopia	July to September	March to May	October to January
Region B: western Ethiopia	March to November		December to February
Region C: southern and south-eastern Ethiopia	March to May	September to November	December to February and June to August

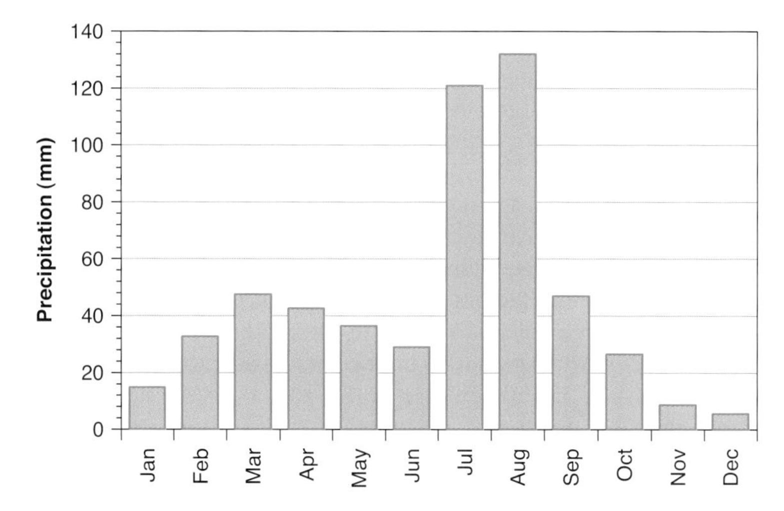

Figure 4 Mean monthly precipitation recorded at Matahara (1966–2004)

2.1 Precipitation

The Upper Awash System has two precipitation seasons: the March to May short rainy season and the July to September long rainy season, which are separated by a relatively short dry period in June. Table 1 gives a summary of the seasons of the three climatic regions of Ethiopia. Figure 4 shows the mean monthly precipitation record at Matahara town for the 1966 to 2004 historical period.

Average annual precipitation at Matahara is about 543 mm, and it varies from about 864 mm in wet years to 310 mm in dry years. The precipitation in the region is highly erratic, resulting in a very high risk of annual droughts and intra-seasonal dry spells.

Frequent delays and shortfalls of seasonal precipitation seriously affect water supply management in arid and semi-arid regions of Ethiopia. Some indications of decline in precipitation have been observed in this region and other regions of the country since 1997, as shown in Figure 5. The greatest percentage decrease has occurred during the February to May season. The July to September season rains have also diminished in many areas. This recent dryness may be linked to a warming tendency in the southern Indian Ocean, and is likely to affect water-stressed regions of arid and semi-arid regions of Ethiopia.

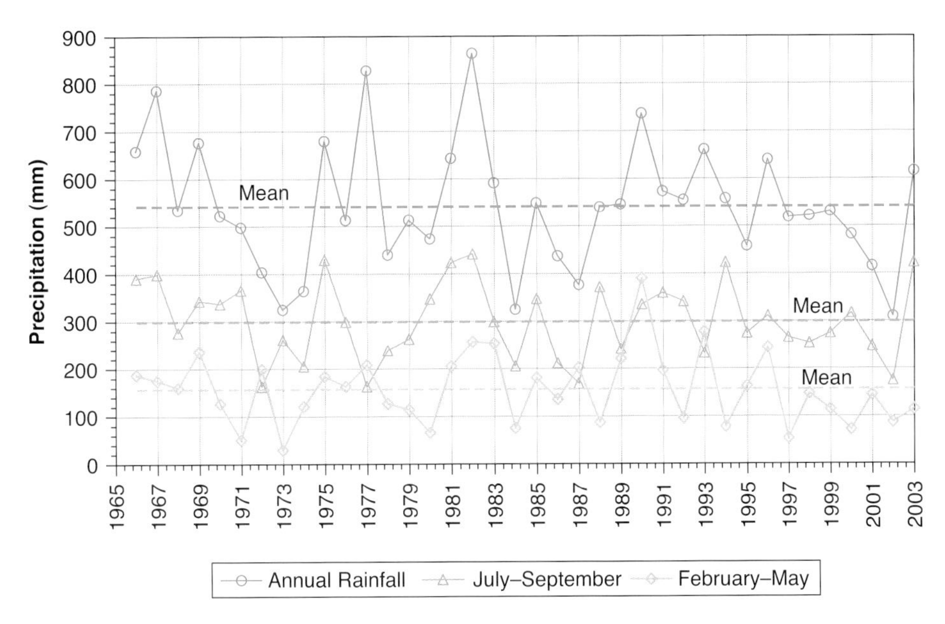

Figure 5 Annual and seasonal variation of precipitation recorded at Matahara

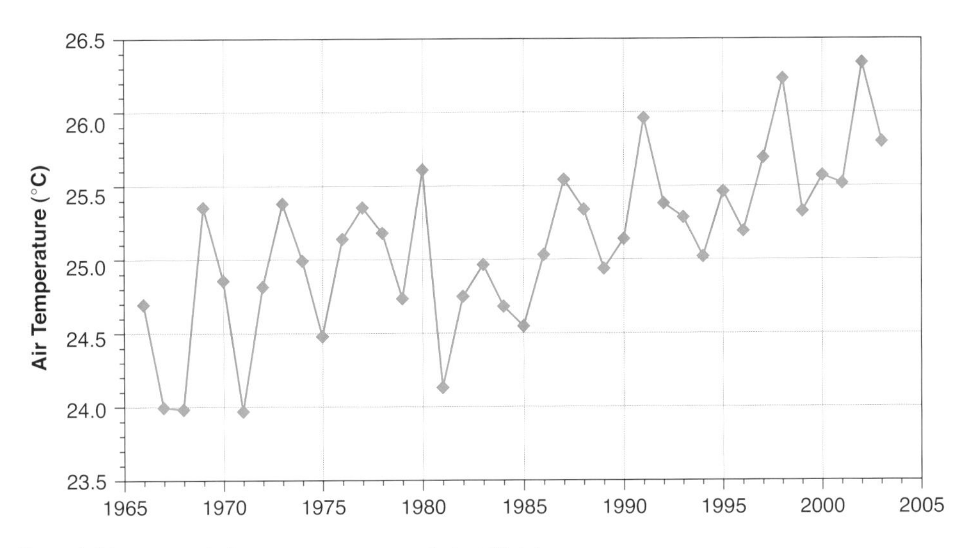

Figure 6 Mean annual air temperature recorded at Matahara

2.2 Temperature

The temporal distribution of mean annual air temperatures derived, based on recorded data at Matahara climate station, are shown in Figure 6. The recorded data indicates that the mean annual air temperature varies over a narrow range of about 24°C to 26.3°C. However, the recorded mean annual air temperature shows an increasing trend since 1990.

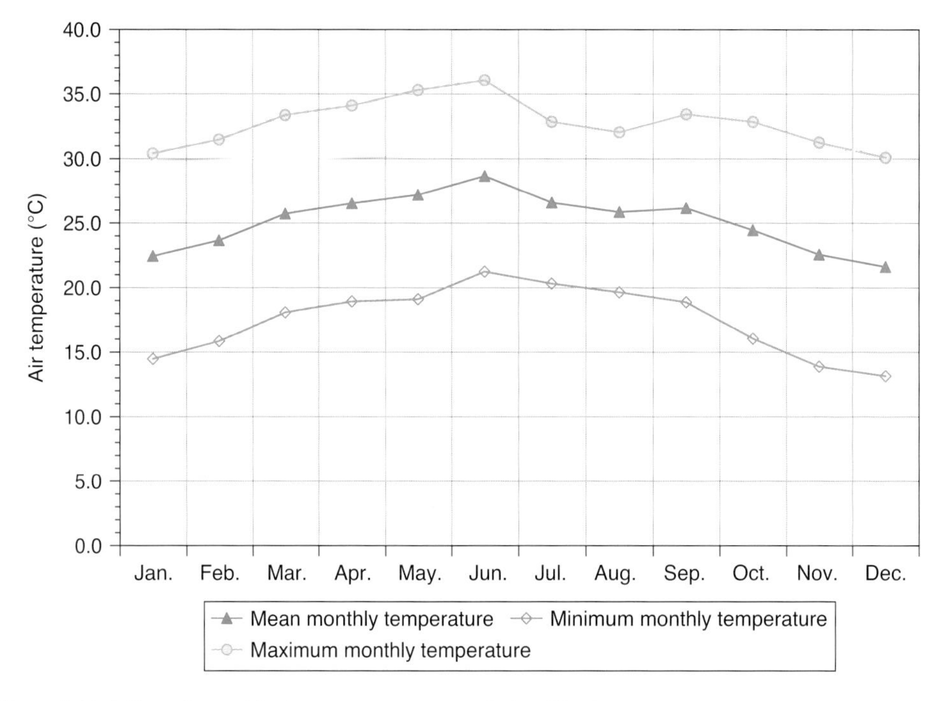

Figure 7 Variation of monthly air temperature recorded at Matahara

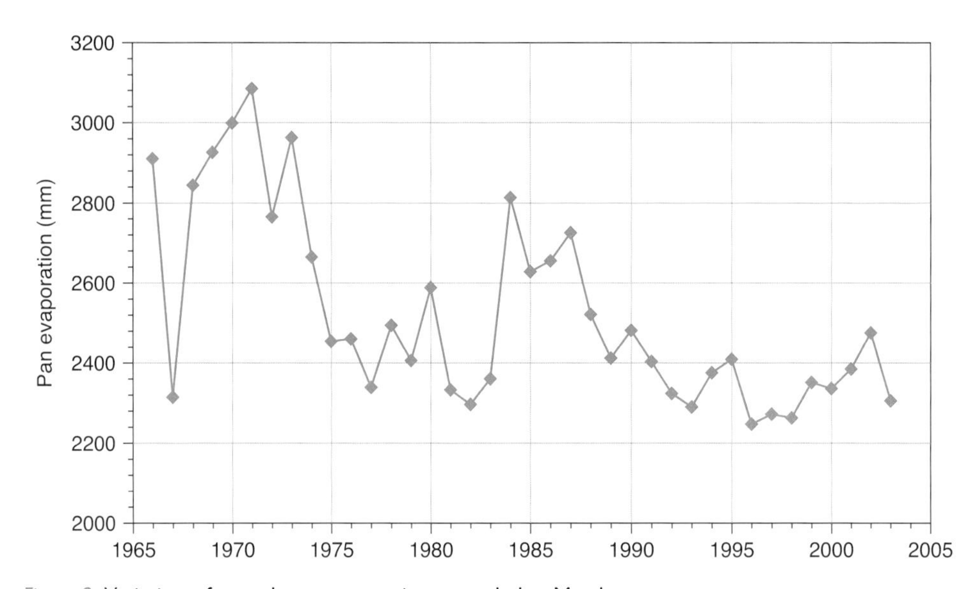

Figure 8 Variation of annual pan evaporation recorded at Matahara

The mean monthly temperatures typically ranged from 28.6°C in the month of July to 21.6°C in December, as shown in Figure 7.

2.3 Evapotranspiration

Mean annual potential evapotranspiration varies between 1,700 and 2,600 mm in arid and semi-arid areas of Ethiopia. Recorded pan evaporation data at the Matahara climate station indicate that annual pan evaporation varies from 2,250 mm to 3,080 mm, as shown in Figure 8. The recorded pan evaporation data indicates a decreasing trend in pan evaporation since 1990.

3 GEOMORPHOLOGIC CHARACTERISTICS

The Great Rift Valley is a geological feature that runs north to south from northern Syria in south-east Asia to Central Mozambique in East Africa. This geological feature is the result of the rifting and separation of the African and Arabian tectonic plates that began some 35 million years ago in the north, and by the ongoing separation of East Africa from the rest of Africa along the East African rift, which began about 15 million years ago during the Miocene.

The geology of the Middle Awash region is characterized by major stages of floor subsidence since the Miocene, and active volcano of either basaltic or acidic nature. This active volcano is often associated with differentiated magma that is manifest in a geothermal anomaly of warm springs and hydrothermal alteration. Some of the identified geologic units in the region include aphyric basalt, marshy soil, alluvial deposits, colluvial deposits, ash deposits, porphyritic basalt rhyolite and welded tuff (ELC and Tropics, 2005). Available drilling data, up to a depth of 20 metres in limited locations, also shows localized layers of deposits. According to the ELC and Tropics (2005) study, a 20-metre deep log of fairly weathered basalt, tuff with a sandy mixture and another basalt layer of, presumably, Pleistocene era was found in one location; whereas a loose silty clay and clayey silt with a fair sand content was found for the entire 20 metres depth of drilling in a nearby location.

These features, in both the horizontal and vertical directions, make the transmittivity values in the region highly localized and variable.

4 WATER SUPPLY SOURCES

Rainfall runoff is the main source of water in the region. The mean annual rainfall in the Awash Basin ranges from 200 mm in the north to 1900 mm in the south (Halcrow, 1989). In the catchment area of the basin, the yearly rainfall distribution varies from a minimum of 700 mm to 800 mm in the lower part to more than 1,600 mm on the mountain peaks along the upper basin boundary (ELC and Tropics, 2005). The estimated annual runoff in the basin is about 4.6 km^3 (Halcrow, 1989).

UNDP (2006) reports that, in Ethiopia, rainfall is both highly seasonal and exceptionally variable over time and space. This report points out further that this variability, combined with a limited infrastructure for storage and poorly protected watersheds, exposes millions to the threats of drought and floods.

4.1 Surface water

Koka dam is the main source of surface water supply to the various projects. The catchment area upstream of the dam is about 11,200 km^2. This is about 10% of the total basin area but provides the bulk of water resources for the various projects in the basin, virtually all of which are located downstream of the dam. The other storage reservoirs in the basin are the Kassam Dam[1] on the Kassam River, a tributary to the Awash River, in the Middle Awash system, and the Tandaho Dam in the Lower Awash system.

4.2 Groundwater

There is inadequate exploration of the groundwater resources of the region. It is reported that at one location in the Upper Awash system, the groundwater level is at a depth of about 20 metres below the surface and it is likely that the Awash River may be recharging the groundwater. The Rift Valley's rich salt deposit and active volcano render the groundwater in the area not readily suitable for irrigation or municipal purposes. The ELC and Tropics (2005) report suggests that due to its high temperature, total dissolved solids, electrical conductivity, and sodium and fluoride concentration, the spring water in the region is not suitable for drinking purposes. The high fluoride content of the groundwater markedly affects the bone development of users of the groundwater. This is evident from a permanent discolouration on the teeth of those groundwater users; this groundwater effect is also known as fluorosis or mottling of teeth.

However, the geothermal resource could be a significant source of power for industrial purposes.

5 WATER DEMAND

There is a wide variety of water users in the region. The municipalities in the Upper Awash system include Adama, Wanji and Matahara. There are sugarcane plantations for the Wanji and Matahara sugar factories. The Marti tomato paste and marmalade processing plants are also located in the area. There are also a number of small-scale irrigation schemes along the river, which are operated by the farming community in the system. Hydropower generating plants are also located at the Koka dam, Awash II and Awash III. A fourth generating plant, called Awash IV, is being considered for construction. Besides these water users, sedentary and pastoral communities inhabit the region.

5.1 Municipal

Adama, one of the biggest cities in Ethiopia, gets its water supply directly from the Awash River below Koka dam. The towns of Wanji and Matahara also get their water supplies directly from the Awash River. Based on a 1994 census by the Government of Ethiopia, the projected 2006 populations of Adama and Matahara are nearly 230,000 and 22,000, respectively. Wanji is a smaller town than Matahara, but not unimportant in its sugarcane plantation and sugar factory, which is the oldest in the country.

[1]This dam is under construction and is expected to become operational in the near future.

Latest figures from the World Bank put Ethiopia's estimated mean annual population growth rate at about 1.8% during 2005 (World Bank, 2007). Historically, the urban growth rate has been significantly higher, partly because of a net rural to urban population migration trend. There are also signs of a significant economic growth rate in the country in recent years. The World Bank's estimated Gross Domestic Product (GDP) growth rate for Ethiopia is 8.7% during the year 2005.

According to an August 2005 report by the Ethiopian Water Works Design and Supervision Enterprise (WWDSE, 2005), the estimated current annual diversion for the city of Adama is about 12 million cubic metres.

5.2 Industrial

The Matahara and Wanji sugar factories are the two oldest factories of their kind in the country and provide the bulk of the country's sugar consumption. In recent years these factories have started to export their products to markets outside Ethiopia. This trend suggests that the water demand of these factories, along with their sugar plantations, is likely to grow significantly. Other factories include the Marti tomato paste and marmalade processing plants. These factories are conveniently located near the Awash River bank and get their water needs by way of direct weir diversion and pumping from the river. Because of the growing demand for sugar in the domestic and foreign markets, it is believed that the Ethiopian government will expand the sugarcane plantations and the sugar factories.

Floriculture enterprises are emerging in the upper reach of the basin. With a lucrative flower export trade to European markets and the latest reported relocation of such enterprises from other African countries to Ethiopia, this system is more likely to see the expansion of this kind of enterprise in its reach in the future.

5.3 Irrigation

According to a 1993 UNDP report, the Awash River Basin has most of the irrigation developed to date in Ethiopia (UNDP, 1993). We believe that this trend has not changed since then. FAO (1997) estimates that the total water requirement in the Rift Valley is about one-quarter of the available runoff. This report indicates that the irrigation potential of the Awash Basin in general is 205,400 hectares.

The irrigation water requirement in the region ranges between 5,000 and 10,000 cubic metres per hectare per year (FAO, 1997). The sugar-processing plants mentioned earlier depend on vast sugarcane plantations. Tomato fields as well as irrigation fields at Bofa and Walanchiti are also located in the area. All these fields get their irrigation water from the Awash River. Future expansions and any endeavour for the utilization of the irrigation potential would require important storage works.

5.4 Recreational

The Sodaré resort is an important tourist spot bordering on the Awash River. It is located a little over a hundred kilometres from the capital and at about 12 km from the city of Adama. The Awash National Park is also found farther downstream at about 225 km to the east from the capital and bordering on the Awash River. This park is the oldest and most developed wildlife reserve in Ethiopia. The park is host to many East

African plain animals and about 450 species of birds within its 720 km^2 field (Tadesse, G. et al., undated).

6 INTEGRATED PLANNING AND MANAGEMENT

The Upper Awash System is a functionally stretched system. It is affected by the quantity and quality of the water it receives from upstream. It is affected by agricultural, municipal, industrial, hydropower, recreational and pastoral water demands. Downstream there are irrigation water supply requirements. The fact that the hydropower generated from the Awash River is connected to the national hydropower grid makes its management dependent on variables outside this system.

The various agricultural, municipal and industrial water projects were built, and continue to be built, in a piecemeal fashion. There are numerous industrial activities in the Upper Awash System. Addis Ababa, Ethiopia's capital with a population of nearly 3 million, is located within the Upper Awash System and its metropolis has a highly disproportionate number of the factories in the country. Since Addis Ababa is in the Awash River Basin, the Awash River bears the burden of the country's potential point sources of pollution.

The Middle and Lower Awash Systems have a series of irrigation activities with pervasive salinity problems due to the rise in groundwater level. These problems need an expensive drainage system for leaching purposes. In fact, earlier projects in this region resulted in poor drainage networks that were upgraded at very high costs. The operation of these projects has been found to be problematic due to poor coordination in transferring management responsibility from the federal government to the local state government after the upgrading of the project was completed. The Amibara subsurface drainage project is a case in point.

6.1 Water supply

The water supplies for the municipal, agricultural, industrial, hydropower and recreational purposes are mainly provided by government agencies. In 1998, the central government of Ethiopia instituted the Awash Basin Water Resources Administration Agency (ABWRAA). Its purpose is coordinating, administering, allocating and regulating the utilization of the surface water resources of the Awash River Basin. It is responsible for collecting water resources data, planning, designing, implementing and overseeing operation of all water projects within the basin (Achamyeleh, undated).

The practical significance of this young institution is yet to be observed. However, environmental impact assessments of new project initiatives are taking root in newly initiated projects. Regulatory requirements are not clear enough at this time.

The country's water law is yet to be well understood. Interstate competition between Oromia and Afar states as well as state and central government approaches to regulatory requirements have been observed in recent times. Edossa (2005) points out differences in Ethiopia's statutory systems and customary systems of water resources utilization. He argues that the statutory systems use a top-down approach to regulate. This problem has manifested itself in the differences between Oromia state's concerns to impose the central government's statutory requirements on the directly affected users.

A more recent report indicates that Ethiopia's decentralization effort has transferred a high level of authority to district and village level bodies though financial and human capacities remain weak, and in some areas the legal status of village water supply and sanitation committees is not recognized (UNDP, 2006).

In the future, these factors are bound to play more important roles in the planning, management and operation of the region's water resources.

6.2 Water quality

The fluvial soil characteristics of the high runoff region of the Awash River Basin, the disproportionate concentration of the country's industry in the basin and the active volcano in the Middle Awash System, make its fresh water resources burdened with both point source and distributed source pollutants. The average annual soil erosion in the basin's catchment area is around 200 to 300 tons per hectare (Halcrow, 1989). According to a 2002 study by the Ethiopian Electric Power Corporation, the sediment-laden inflow to Koka Dam has already reduced its storage capacity by 30% (EEPCO, 2002). In fact, its designed sediment outlet work has already fallen below the surface of the settled sediment.

Tadesse (undated) reports that, in the Middle Awash, evapotranspiration exceeds mean annual rainfall, which leads to the accumulation of groundwater salts on the surface.

There are insufficient, practical water quality monitoring systems in the country, in general, and the Awash River Basin in particular.

6.3 Water treatment

Ethiopia's national drinking-water quality standards follow the World Health Organization's (WHO) guidelines. The city of Adama receives treated water that meets WHO's standards. The water supply for Adama is pumped from the Awash River to a treatment facility that consists of sedimentation basins, sand filters, and other biological

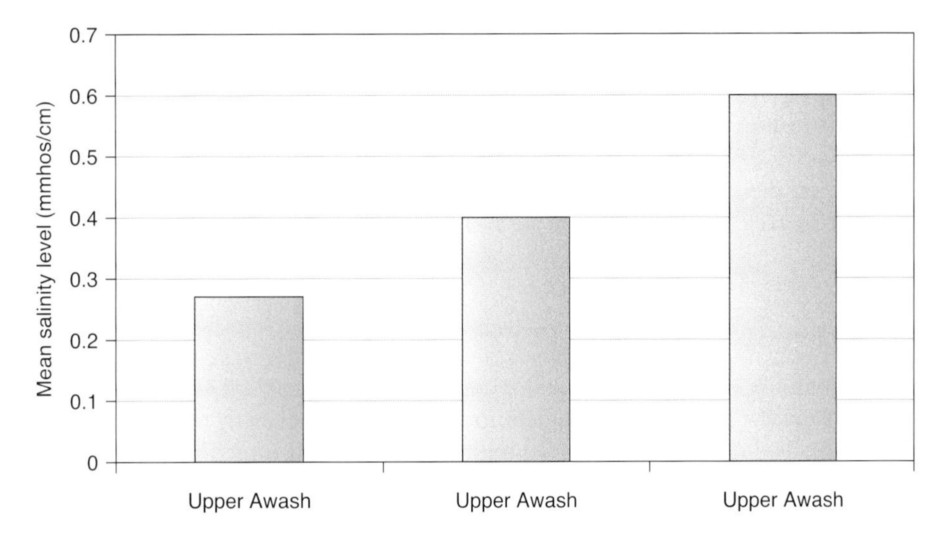

Figure 9 Mean salinity level of Awash River water

Source: Halcrow, 1989

and chemical treatment facilities. The treated water is stored in service reservoirs and delivered to the distribution system by gravity.

However, the town of Wanji and other small communities in the surrounding areas use untreated groundwater for domestic purposes. The groundwater has a high fluorine content, which particularly affects these communities.

6.4 Sewer management

As a developing country, Ethiopia faces a vast sanitation problem. According to the United Nations Development Program's Human Development Report of 2006, 1 in 2 people in developing countries lack access to improved sanitation (UNDP, 2006). This report shows that, for Ethiopia, the national average of people with access to sanitation stands at about 13%. The rural–urban divide of this average figure stands at less than 10% and nearly 45%, respectively. This average figure shows that only about 1 in 7 people have access to improved sanitation. The specific figure for the Upper Awash System is likely to be higher than the national average because of better developmental activities in the region. Nonetheless, it is far below an acceptable level.

The cities of the Upper Awash System find themselves in this predicament. Sewer treatment and disposal are some of these cities' very crucial problems. Treatment of wastewater from public sewers is practically absent. Water-related diseases such as diarrhoea and parasitic infections continue to be some of the problems.

Although septic tanks are used by public institutions such as hospitals, schools and office buildings, conventional wastewater collection and treatment facilities are absent. Pit latrines are commonly used wastewater disposal facilities for households. Recently, DHV Consultants undertook a feasibility study of a Wastewater Master Plan for Adama, which is the first of its kind dealing with integrated wastewater for this city.

6.5 Excess water management

Ethiopia's plateau generally has heavy rain during the June to August season, making flooding from runoff a frequent concern. River and levee overflow are not uncommon through the entire Awash River course. The city of Adama lies at the foot of north, east and west escarpments. During the high rainfall season, the runoff from these escarpments flows to the city and causes significant flooding. Other than roadside storm drain gutters, there are no stormwater protection facilities to mitigate the problems. These gutters are often used as a disposal outlet for some household wastewater.

7 SYSTEM SUSTAINABILITY

The sustainability of the Upper Awash System requires important consideration by all the parties involved. Government institutions that can deal with this issue are being put in place. The ABWRAA is a good initiative in the right direction. However, comprehensive studies that assess system sustainability are lacking. The following subsections suggest some helpful analyses of water resources constraints, and planning and management options for system sustainability.

7.1 Drought management

In general, Ethiopia is better endowed with water than many drought-prone countries. The problem of water shortage emanates from the seasonality of rainfall and the lack of infrastructure for storage to capture excess runoff during flood seasons. UNDP (2006) reports that rainfall variability is estimated to have pushed an additional 12 million people below the absolute poverty line in the second half of the 1990s. This report notes that Ethiopia stores 43 cubic metres of water per person out of the per capita water availability of 1,644 cubic metres per person. This shows a mere 2.62% utilization of the potentially available water resource for every person in the country. The Awash basin in general, and the Upper Awash system in particular, may not have significantly different indicators in this regard. The UNDP report (2006) concludes that Ethiopia's water problem is predictability rather than availability.

It may be necessary to consider various drought management strategies that have been tried in other countries. Dziegielewski et al. (1996) lists the following drought management options in response to anticipated shortages of water: 1) demand reduction options, 2) improvements in efficiency in water supply and distribution system, and 3) emergency water supplies. An institutionalized characterization of droughts and a capability for foresight are important steps for the management of potential droughts.

7.2 Mitigation of impacts of climate variability

Various studies have indicated that the precipitation variability in Ethiopia is closely related to the fluctuation of the Southern Oscillation (SO), which is triggered by Sea Surface Temperature (SST) anomalies during El Niño (warm SST) and La Niña (cold SST) events. These fluctuations in atmospheric circulation coupled with ocean phenomena have significant impacts on the position, magnitude and intensity of rain-bearing systems in Ethiopia.

7.2.1 Effect of ENSO and LNSO events in arid and semi-arid regions of Ethiopia

El Niño is the increase in SST in the central and eastern equatorial Pacific Ocean as a result of changes in the pattern and direction of winds and ocean currents in the region. The phenomena, coupled with changes in atmospheric pressure across the Pacific Basin between Darwin, Australia and Tahiti, is called as the Southern Oscillation (SO). The SO leads to devastating droughts in northern Brazil, Australia and parts of Africa; the failure of the Indian monsoons; and hurricanes along the east coast of North America (Glantz, 1993). However, La Niña, which is a decrease in SST, has a reverse impact including heavy precipitation and flooding in Africa.

The El Niño – Southern Oscillation (ENSO) – and La Niña – Southern Oscillation (LNSO) – episodes are strongly linked with the various atmospheric systems and precipitation distribution over Ethiopia. Table 2 shows the relationship between past extreme ENSO and LNSO events and meteorological drought in Ethiopia.

An ENSO event results in normal and above normal February to May season precipitation and widespread meteorological drought (less and erratic precipitation) during the June to September season. LNSO is strongly associated with the deficiency of precipitation during the former season and heavy precipitation during the latter season.

Table 2 Occurrence of El Niño and La Niña events and its relation to drought and flooding in Ethiopia

Year	*El Niño year*[a]	*La Niña year*[b]	*Ethiopian meteorological condition*
1982/83	+2.3[c]		Severe drought (1983/84)
1986/87	+1.2		Drought (1987/88)
1987/88	+1.6		Drought (1987/88)
1988/89		−1.9	Flooding (long rainy season in 1988)
1991/92	+1.7		Drought (1990–1992)
1994/95	+1.3		Drought
1995/96		−0.8	Flooding (1996)
1997/98	+2.5		Drought (1997/98)
1998/99		−1.5	Localized flooding
1999/2000		−1.6	Localized flooding
2000/2001		−0.7	Localized flooding
2002/2003	+1.5		Drought (2003)
2004/2005	+0.9		Drought (2005 and 2006)

[a] Pacific Ocean Niño Index (ONI) $>$ +0.5°C for a minimum of 5 consecutive over-lapping seasons.
[b] Pacific Ocean Niño Index (ONI) $>$ +0.5°C for a minimum of 5 consecutive over-lapping seasons.
[c] Maximum Oceanic Niño Index (ONI) in the year, where ONI is a three month running mean of ERSST.v2 SST anomalies in the Niño 3.4 region (5°N–5°S, 120°–170°W)
Source: http://www.cpc.ncep.noaa.gov/products/analysis_monitoring/ensostuff/ensoyears.shtml

Based on an analysis of seasonal precipitation from 233 precipitation stations in Ethiopia, Bekele (1992, 1993) established a relationship between the short rainy season and the long rainy season precipitation and eastern equatorial Pacific SST anomalies. The impacts of El Niño and La Niña on the Ethiopian climate system can be preliminarily identified based on the following conditions.

- An El Niño event will result in severe and widespread drought during the long rainy season if the increase in SST is greater than 1°C and lasts at least 12 months. The 1972/73, 1982/83 and 1987/88 El Niño events can be taken as examples.
- A positive SST anomaly is mostly associated with normal and above normal precipitation amounts in the short rainy season.
- A La Niña event with a significant decrease in SST is strongly associated with precipitation deficiency in the short rainy season and flooding in long rainy season. The 1988/89 La Niña event can be taken as an example.

The effects of ENSO and LNSO events on the precipitation regimes of the arid and semi-arid region of Ethiopia are shown in Figure 10. This figure is based on recorded data at the Matahara climate station. The ENSO event in 1982/83 resulted in the reduction of precipitation for the entire year of 1984 in all three regions and resulted in severe droughts throughout Ethiopia. The 1988/89 LNSO event resulted in significant increases in precipitation from June to October in this region.

7.2.2 ENSO and LNSO events and hydrological drought and flooding in Ethiopia

Hydrological drought refers to deficiencies in surface and subsurface water supplies. It is measured as stream flow volume, and lake, reservoir and groundwater levels. Climate is

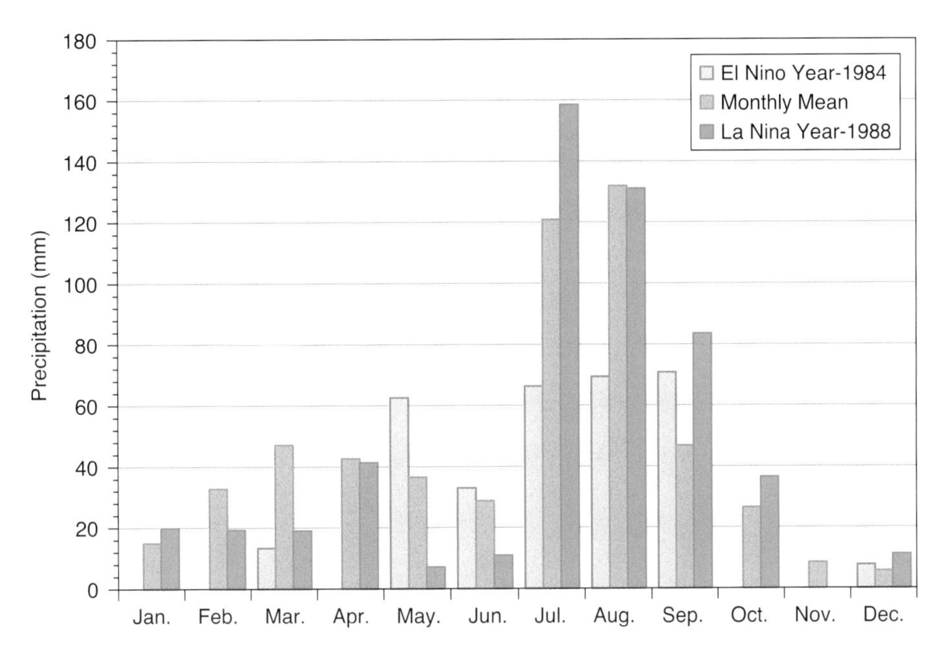

Figure 10 Effects of ENSO and LNSO events on precipitation in arid and semi-arid regions of Ethiopia (See also colour plate 24)

a primary contributor to hydrological drought, although other factors, such as changes in land use (for example, deforestation), land degradation and the construction of dams, will also affect the hydrological characteristics of the basin. Since various watersheds are interconnected by hydrologic systems, the impact of meteorological drought may extend well beyond the borders of the low precipitation areas.

The effect of ENSO and LNSO events on hydrological drought in Ethiopia are similar to meteorological droughts, as shown in Figure 11. The ENSO events resulted in hydrological drought in 1992 in all regions of Ethiopia. The LNSO events resulted in excessive flooding in the long rainy season and in runoff deficiency in the short rainy season in 1988/89, as shown in Figure 12 for Lake Abaya (located in the Rift Valley region).

7.2.3 Use of ENSO and LNSO information for water supply management

Water supply and demand are considerably more sensitive to climate variability because of ENSO and LNSO events in the arid and semi-arid regions of Ethiopia. For example, water demand per capita from storage reservoirs may increase during an ENSO event since input from precipitation will be reduced significantly. However, extreme flood events associated with LNSO events can result in excessive damage to water supply infrastructures if the events are not taken into account during the design of the system.

The relationship between ENSO and LNSO events to meteorological and hydrological phenomena in Ethiopia suggests that general forecasts of the effects of these events can be made several months in advance and the result of the analysis can be incorporated into water supply management and operation (Biftu and Sawatsky, 2004).

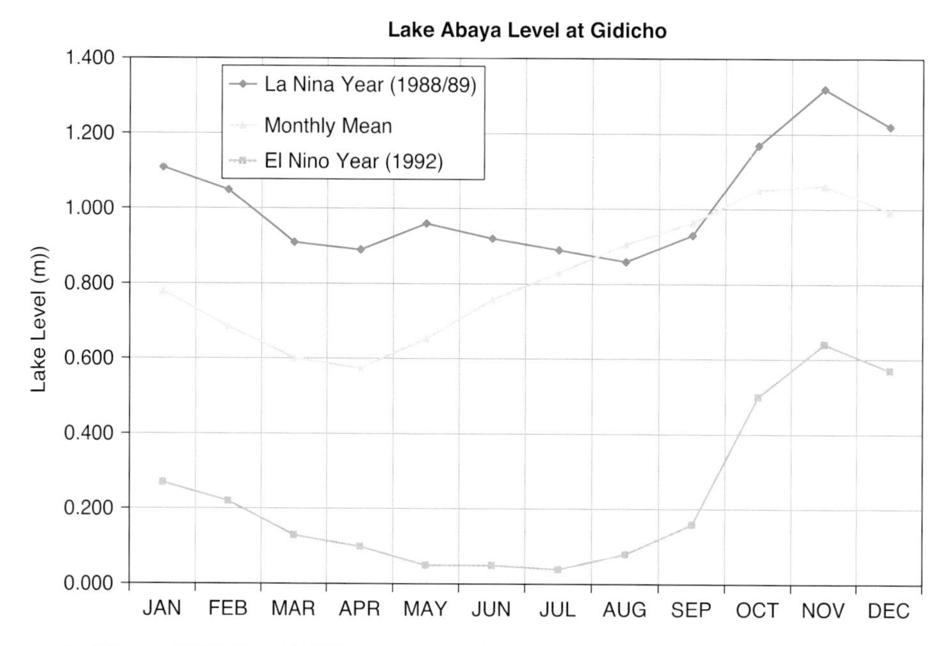

Figure 11 Effects of ENSO and LNSO events on hydrologic regime in Ethiopia

At the planning and design stages of urban water supply systems, a more robust and integrated water supply management plan can be developed, based on past information of extreme ENSO and LNSO events. The plan may include:

- Design of water supply infrastructure for long-term sustainability based on an understanding of the cyclical nature of the climate variability in the region.
- Use of more flexible water allocations that do not permanently grant fixed amounts of water to users that may not be sustainable in the future.
- Establish temporary change in operation rules for surface water reservoirs, based on expected changes in precipitation and runoff forecast taking into account the effect of ENSO or LNSO events.
- Include mitigation and adaptation measures during operation of the water supply systems based on ENSO and LNSO forecast information.

The most effective mitigation strategy is to prepare, issue and publicize timely ENSO and LNSO related forecasts. If properly warned, the public can take action to minimize the impacts of a potential disaster as a result of drought or flooding.

7.3 Impacts of potential global warming

In the past, water planning and management relied on the assumption that the future climate would be the same as the past. However, there is now a broad consensus that global warming is a real problem and that it will alter the hydrologic cycle in a variety of important ways (IPCC, 2001). Yet, there is little certainty about the form these changes

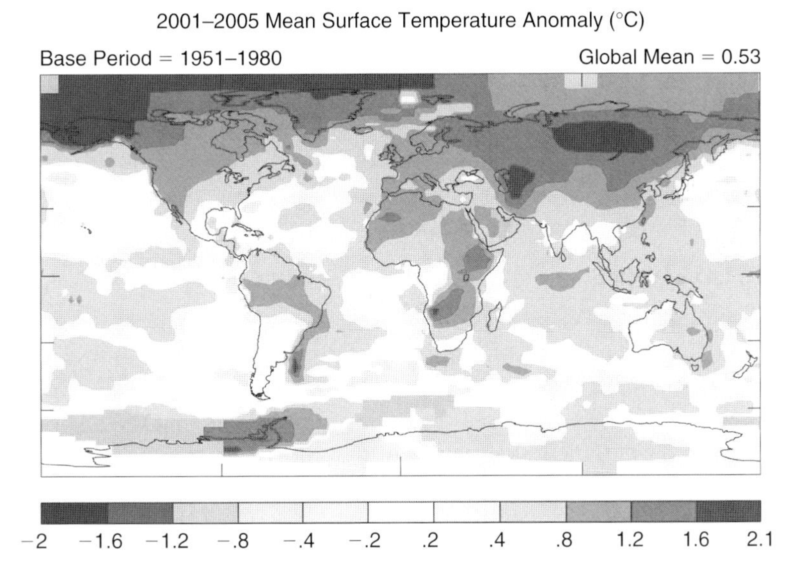

Figure 12 Global mean surface temperature anomaly (See also colour plate 25)

Source: Hansen, 2006

will take, or when they will be unambiguously detected. Despite the uncertainties and debates about the effect of potential climate change, more attention is being paid to possible consequences of climate change. This is particularly true in the area of water resources managements where many decisions depend explicitly on the assumptions we make about future climatic conditions.

Among the potential impacts of climatic changes are higher regional temperatures, decreases in average precipitation, changes in the regional patterns of precipitation, changes in the precipitation intensity, changes in the timing of major storms and increased regional evaporation. These changes are bound to alter both the demand for and the quantity and quality of water supply in the arid and semi-arid regions of Ethiopia.

The impact of climate change on the water resources of the Upper Awash is yet to be inferred. Some analysis of the results from some global General Circulation Models (GCMs) for a recent historical period, shows climate anomalies in the region on the higher side of the global mean. Figure 12 shows that for the Upper Awash System, the mean surface temperature increase during the 2001–2005 period, compared to the 1951–1980 base period is about 1.4 degrees Celsius (Hansen, 2006). The corresponding global anomaly is estimated to be about 0.53 degrees Celsius. These figures suggest that according to this particular study, the region of this case study has experienced a nearly three-fold surface air temperature increase compared to the global mean anomaly.

Therefore, Matahara, Wanji and Adama will probably have significant moisture deficit, and be vulnerable to any long-term changes in climate. Although our present state of knowledge does not allow predictions of the direction of climate change or climate variability in this region with confidence, there are indications from observed data and climate models that the region may become drier and warmer, with increases in air temperature and seasonal variation in precipitation distribution.

In summary, in the future, climatological variability, compounded with the effect of climate change, will have direct and indirect impacts on a wide range of social, economic and environmental aspects. Although the quantification of the impact is lacking, the water management of the system is likely to be more challenging in the future. Therefore, current water supply management policies and tools need to be enhanced to assess the impacts of climate change on water supply, demand, quality and the environment.

8 CONCLUSIONS

The Upper Awash River System is a highly constrained system due to projects and development activities from within, upstream and downstream. Ethiopia's significant economic growth tendency in recent years is likely to attract more investment opportunities in the near future and there is little reason to believe that the disproportionate concentration of developmental activities in the upper region of the basin is going to ease anytime soon. Industrial, agro-industrial and floricultural enterprises are making considerable emergence, particularly in the upper region of the basin. With all these activities, there is reason to believe that water resources demand in the system will increase significantly. Furthermore, distributed and point source pollution will be one of the major concerns in the system.

Ethiopia's relatively high average population growth rate and even significantly higher urban population growth rate are going to further constrain the management of its water resources. The present per capita water use is likely to grow if the current economic growth trend of the country manages to continue. This is particularly important in the Awash River Basin, in general, and the Upper Awash System, in particular.

The poor infrastructure to capture and store surface runoff during the short rainy seasons, the effect of climate variability and climate change are formidable factors to deal with for a sustainable management of the water resources in the system. In a system where the predictability of water is far more problematic than its availability, integrated management of this resource is imperative. In addition, stormwater and sewer management in the Adama, Wanji and Matahara areas remain at a rudimentary stage.

The institutional capacity for integrated water resources planning and management in the basin, in general, and in the system, in particular, are emerging. Interstate relations, local government to central government relations and interagency cooperation will be necessary to manage the system in a sustainable way. There is an excellent opportunity for integrating the planning, management and operation of all the projects in the Awash River Basin. Planning according to projected water demand 20 to 30 years into the future is also necessary.

ACKNOWLEDGEMENTS

We would like to acknowledge the National Meteorological Services of the Ethiopian Ministry of Water Resources as well as Dr. Seleshi Bekele of the International Watershed Management Institute in Addis Ababa for providing meteorological data of the study area.

REFERENCES

Achamyeleh, Kefyalew, undated. *Ethiopia: Integrated Flood Management.*

Bekele, F. 1992. *The Effect of Southwest Indian Ocean Cyclones on Belg and Kiremt.* NMSA mimeo. Addis Ababa, Ethiopia: NMSA.

Bekele, F. 1993. Ethiopian Use of ENSO Information in its Seasonal Forecasts. M.H. Glantz (ed.) *Workshop on Usable Science: Food Security, Early Warning and El Niño.* 25–28 October, 1993. Budapest, Hungary, Boulder: National Center for Atmospheric Research, pp. 117–121.

Biftu, G.F. and Sawatsky, L.F. 2004. Impact of Climate Variability on Small Scale Irrigation Systems in Drought Prone Areas of Ethiopia. *57th Canadian Water Resources Association Annual Congress, Water and Climate Change: Knowledge for Better Adaptation,* June 16–18, 2004, Montreal, QC, Canada.

Dziegielewski, B., et. al. 1996. *Drought Management Options, Drought Management and its Impact on Public Water Systems.* Report on a colloquium sponsored by the Water Science and Technology Board, Sept. 5, 1985, National Academy Press, Washington, D.C.

Edossa, D.C., et al. 2005. Indigenous Systems of Conflict Resolution in Oromia, Ethiopia. *African Water Laws: Plural Legislative Frameworks for Rural Water Management in Africa,* 26–28 January 2005, Johannesburg, South Africa.

ELC and Tropics. 2005. *Awash IV Hydropower Project Feasibility Study, Design Works and Tender Document Preparation, Inception Report for the Ethiopian Electric Power Corporation,* February 2005.

Ethiopian Electric Power Corporation. 2002. *Koka Dam Sedimentation Study: Recommendations Report,* June 2002, Addis Ababa.

Food and Agriculture Organization of the United Nations. 1964. *Survey of Awash River Basin, a Report to the Imperial Ethiopian Government.* Rome, Italy.

Food and Agriculture Organization of the United Nations. 1997. *Irrigation Potential in Africa: A Basin Approach.* Rome, Italy.

Glantz, M.H. 1993. Introduction. Forecasting El Niño: Science's Gift to the 21st Century. M.H. Glantz (ed.) *Workshop on Usable Science: Food Security, Early Warning and El Niño.* 25–28 October, Budapest, Hungary, Boulder: National Center for Atmospheric Research, pp. 3–11.

Hailemariam, K. 1999. Impact of Climate Change on the Water Resources of Awash River Basin, Ethiopia, *Climate Research,* pp. 91–96.

Hansen, J.E. 2006. *Global Warming: Is There Still Time to Avoid Disastrous Human Made Climate Change?* Discussion on 26 April 2006 by Jim Hansen, National Academy of Sciences, Washington, D.C.

IPCC 2001. Third Assessment Report – Climate Change 2001, *The Third Assessment Report of the Intergovernmental Panel on Climate Change.* IPCC/WMO/UNEP.

Halcrow. 1989. *Master Plan for the Development of Surface Water Resources in the Awash Basin.* Ethiopian Valleys Development Studies Authority, Final Report, Volume 6.

Tadesse, G. undated. *Evaluation of Water Quality in Middle Awash Valley, Ethiopia.* International Livestock Research Institute, Addis Ababa, Ethiopia.

Tadesse, G., et al. undated. *The Water of the Awash River Basin a Future Challenge to Ethiopia.* International Livestock Research Institute, Addis Ababa.

United Nations Development Program. 2006. *Human Development Report.*

United Nations Development Program. 1993. *Fifth Cycle Country Programme, Sub-programme 2B: Utilisation of Natural Resources for Balanced Development,* Component 2: Water Resources Development, First Draft. Addis Ababa.

Water Works Design and Supervision Enterprise 2005. *Wolenchiti and Bofa Irrigation and Sugar Project.* Draft Final Hydrology Report, August 2005, Addis Ababa, Ethiopia.

World Bank 2007. *World Development Report 2007.* Washington, D.C.

Case Study IV

Water treatment for urban water management in China

Jun Ma, Xiaohong Guan and Liqiu Zhang

School of Municipal and Environmental Engineering, Harbin Institute of Technology, Harbin, China

1 INTRODUCTION TO URBAN WATER SYSTEM IN CHINA

1.1 Current status of water resources in China

China is a large continental country with large water resource endowments of 2,800 billion m^3 or a water resource of 291.7km^3 per km^2 (continent), which is only slightly lower than the global average water resource level of 315.1 km^3 per km^2. However, due to China's large population, the average water resource per capita is only approximately one-quarter of the global average level (Feng et al., 2007). Therefore, China is regarded by the United Nations as one of the thirteen water-deficient countries in the world: there are 300 cities suffering from water shortage, among which 110 cities are extremely short of water, mainly located in north-east China, north-west China and coastal regions. The severe shortage of water resources has become an important factor that restricts the economic and social development of China.

China's water resources are unevenly distributed, as demonstrated in Figure 1. Northern China has only about 20% of the total water resources in China, but it supports more than half of the total population. As a result, the water availability per capita in Northern China is as little as 271 m^3 or one-eighth of the national level and 1/25 of the world average value. Furthermore, the rapid economic development in this region has been extracting significant amounts of water from the environment, and it is also discharging pollution to the water supply sources, which further contributes to water scarcity. In addition, the water availability in northern China varies with seasons, e.g., more than 50% of runoff volume occurs in the summer, which makes the water scarcity problem even more serious.

Over 80% of the water resources in China are concentrated in the south-eastern part of China where the arable area is only 35% of the country's total arable area of 95 million ha. Although southern China is abundant in water resources, some of them are polluted and eutrophicated. These waterbodies normally have certain concentrations of nitrogen and phosphate, due to the discharge of urban and industrial wastewater and agricultural runoff to them (Dokulil et al., 2000). Taihu Lake, West Lake, Chao Lake and Dian Lake are the representative waterbodies that are suffering from eutrophication (Shen et al., 2000; Xu et al., 2005). Eutrophication causes excessive growth of undesired algae and aquatic plants and the excessive algal production can severely deteriorate the water quality in terms of unpleasant odour, bad flavour, high turbidity and low dissolved oxygen concentration.

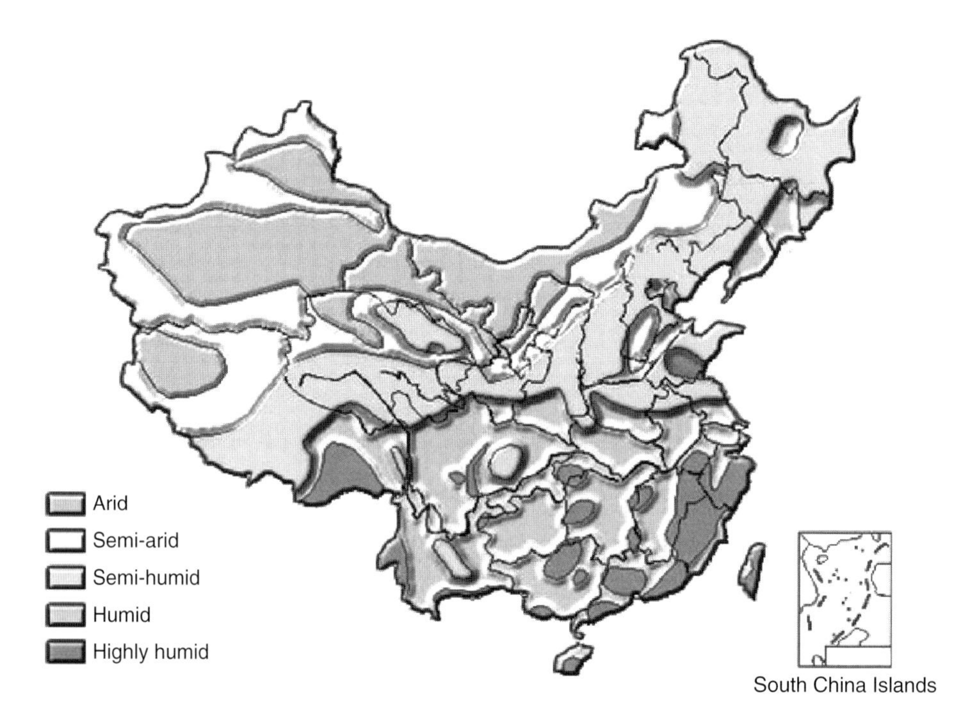

Figure 1 Distribution of water resources in China (See also colour plate 26)

Based on the water quality information of over 1,300 rivers in 2004, it was found that only 59.4% of these rivers could meet or exceed Grade III of the Environmental Quality Standard for Surface Water and 27.8% of these rivers belonged to or were inferior to Grade V. Three hundred and twenty-two reservoirs were assessed in 2004 and 17.7% of them were inferior to Grade V of the Environmental Quality Standard for Surface Water. Among 50 lakes evaluated, 19 of them were substantially polluted and 32 of them were eutrophicated.

1.2 Urban water management in China

The cities in China, especially those in the arid and semi-arid regions, use mainly surface water, including lakes, rivers and reservoirs, as their water resources. If the water quality of water resources is good enough, the conventional treatment process is employed in drinking-water production. However, modified conventional treatment processes or advanced treatment processes are adopted in drinking-water production for the polluted water resource. The drinking water is supplied to the residents and the wastewater is directly discharged to the sewage collection system. For the drinking water supplied to industry, part of the used water is recycled and reused and the rest is treated to comply with certain standard before it is discharged to the sewage collection system.

Both combined and separate sewerage systems are provided to cope with the collection of stormwater and sewage in China. Most of the cities in the arid and semi-arid regions of

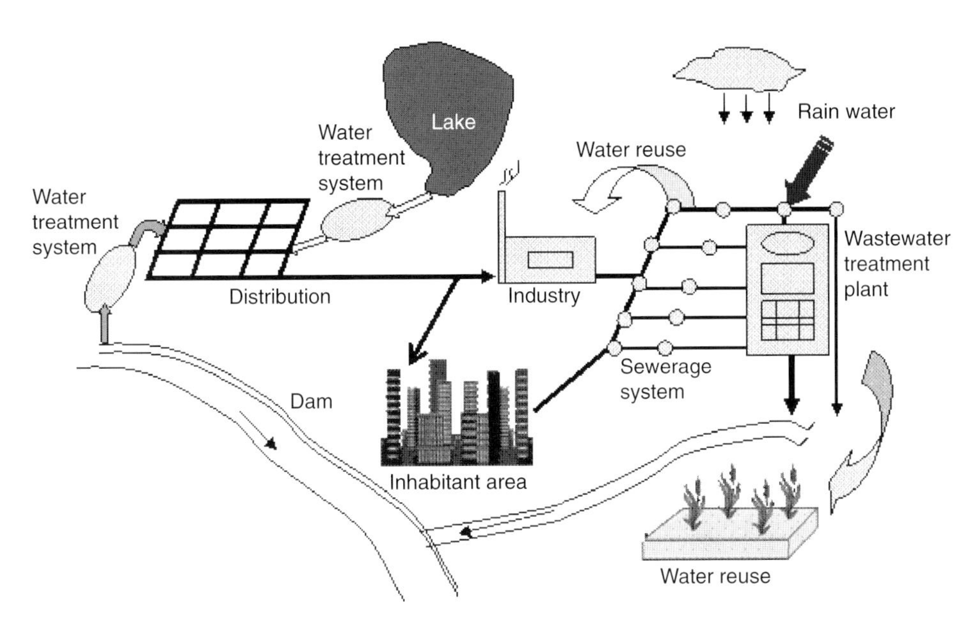

Figure 2 Urban water management system in China (See also colour plate 27)

China have combined sewerage systems as they are cheaper than separate sewerage systems and the precipitation is low in these regions. In the past years, the combined systems were considered to cause high pollution and hygiene risks, while the separate systems were believed to be superior to the combined systems. Therefore, the newly constructed sewerage systems are generally separate sewerage systems. However, in recent years, the discharge from separate sewerage systems has been found to cause adverse effects, such as a non-point source of pollution in lakes and rivers at receiving waters.

Flood control capacity of China is relatively low and flood damages are serious, although China has constructed 277,000 kilometres of river dikes and 85,000 reservoirs. China is one of the countries hit most by severe flood disasters. With fast socio-economic development, more economic stock and higher population density, the flood-prone areas have substantially increased. Flood risk becomes higher and flood disaster mitigation is more important but difficult. According to statistics, since 1990, national average losses resulting from floods were 110 billion Chinese Yuan each year, equivalent to 1% of the national GDP in the same period. The urban water management system in China is illustrated in Figure 2.

1.3 Introduction of water and wastewater treatment in China

In the arid and semi-arid regions of China, the urban areas employ mainly surface water as its source of water supply. Conventional surface water treatment processes, such as coagulation, sedimentation and filtration are commonly used for the water supply. The typical conventional water treatment process is shown schematically in Figure 3.

In certain cases, the enhanced treatment processes such as preoxidation with permanganate, adsorption with powdered activated carbon, dissolved air flotation and biological

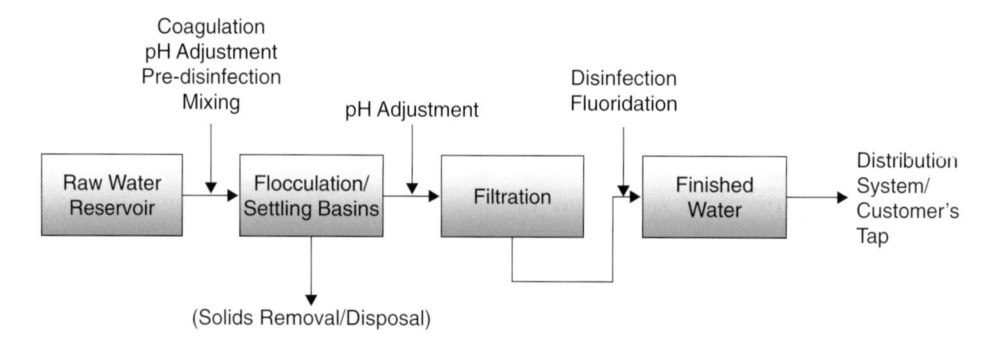

Figure 3 Typical conventional water treatment process adopted in China

filtration are used for improving water quality when raw water is polluted (Ma and Li, 1993; Ma and Graham, 1996; Ma et al., 1994; 1997; 2001). In particular, a composite chemical containing permanganate has been applied successfully to control the taste and odour, to reduce the colour, to control biological re-growth in the distribution system, and to remove iron and manganese. In addition, the permanganate composite chemical (PPC) can enhance the coagulation and filtration processes (Ma et al., 1997) and control the formation of trihalomethanes and other disinfection by-products (DBPs). The combined process of preoxidation with PPC and biological activated carbon has been used successfully in several cities such as Beijing and Jiaxing at relatively low cost for improving drinking-water quality.

As the National Drinking Water Standard becomes more stringent and the water resources are polluted, advanced treatment processes have been developed to remove organic micropollutants. Thus, ozonation, catalytic ozonation and adsorption with granular activated carbon, or the combined processes of preoxidation and biological filtration etc., are employed to modify the conventional water treatment processes to ensure drinking-water safety. Ozone, one of the most powerful oxidants, is effective in inactivating bacteria, viruses and certain forms of algae. Ozone can be used for many different purposes such as: disinfection and algae control, taste, odour and colour control, oxidation of inorganic pollutants (iron, manganese), oxidation of organic micro- and macropollutants and the improvement of coagulation (Langlais et al., 1991). Although ozone is known to be a powerful oxidant, it reacts slowly with some organic compounds such as inactivated aromatics. Moreover, in many cases, it does not cause the complete oxidation of organic compounds (e.g., natural organic matter (NOM)), which results in the formation of biodegradable organic matter (carboxylic acids, carbonyl compounds).

To provide greater oxidation efficiency at low operation cost, catalytic ozonation has been developed (Ma and Graham, 1997; 1999; 2000; Ma and Sui, 2004; 2005). With the presence of a catalyst, the efficiency of ozonation is considerably enhanced, due to the generation of highly oxidative intermediate species. The efficiency of the catalytic ozonation process depends on the surface properties of the catalyst and the pH of the solution, which influences the properties of the surface active sites and ozone decomposition reactions in aqueous solutions. Activated carbon is widely used for the control of synthetic and naturally occurring organic chemicals that are toxic or give undesirable tastes and odours to drinking water. The removal of taste and odour has been one of the initial applications of activated carbons in the water industry with their use now expanded to a wide range of contaminants.

The urban wastewater treatment has made great progress in China in recent years. According to 2005 statistics, there are 708 wastewater treatment plants with a capacity of 49.12 million m^3/d in 661 cities in China. In 2005, 16,280 million m^3 of wastewater was treated, with an urban wastewater treatment ratio up to 45.7% (He, 2006). The treatment capacity and the amount of wastewater treated in 2005 increased by over 100% and 43%, respectively, compared to those in 2000. However, the wastewater treatment capacity cannot satisfy the requirements in most cities and there are still no wastewater treatment plants in 297 cities in China. There are no wastewater treatment facilities at all in more than 50,000 towns and villages.

The wastewater treatment processes adopted in China include the traditional activated sludge process, hydrolysis-aerobic process, oxidation ditch, A/O (Anaerobic/Oxic) process, A^2/O (Anaerobic/Anoxic/Oxic) process and sequential sio-reactor, among which the traditional activated sludge process and oxidation ditch account for 30.6% and 13.9% of the treatment processes, respectively (Wang and Lou, 2004).

In China, there is no special stormwater treatment plant. The stormwater runoff collected by the separate sewerage system is discharged directly to waterbodies. The stormwater collected by the combined sewerage system is transported to the wastewater treatment plant together with sewage to degrade the contaminants contained in stormwater before it is discharged. However, the effluent from these wastewater treatment plants during rainy season cannot meet the standard because the capacity of the wastewater treatment plant is not great enough to treat the mixture of sewage and stormwater. Consequently, some rivers and lakes receiving the treated stormwater and sewage are severely polluted. The authors believe that the rainwater collection and treatment systems in China are not sustainable.

1.4 Summary

According to statistics, 181 cities in China suffered water shortage in 1983. The amount of water shortage was 450 million tons per year. By 1995, the number of cities that were short of water reached 333 (about 50% of Chinese cities) and the amount of water shortage increased to 600 million tons per year (Huang, 1998). In rural areas, about 70.0 million people are in a situation of drinking-water shortage (Yang, 1997). At the same time, water pollution aggravates the water resource shortage. Water resource shortage has become a limiting factor for further development in China (He et al., 2001). The reuse of reclaimed wastewater, including industrial wastewater, municipal wastewater and stormwater has numerous benefits, such as saving fresh water resources, mitigating the conflict of water demand among agriculture, industry and city and alleviating the pollution of water bodies (Tselentis, 1996). The practice of planned water reuse will help to alleviate the serious situation of water shortage in China.

2 URBAN WATER MANAGEMENT IN CHINA – EXAMPLES

2.1 The water resource problems in the urban areas of China

Presently, over 400 of the 660 cities in China are short of water, of which a large proportion results from water quality deterioration caused by pollution. In 2000, nationwide urbanization reached 36%, from 10.6% in 1949. Urbanization in China is expected to be nearly 50% by 2020. In 40 to 50 years' time, urbanization is expected to reach 70%. Rapid urbanization has caused serious water scarcities and drastic conflicts between water demand and supply. Water has become a key limiting factor of the urbanization process, as well as socio-economic development (Varis and Vakkilainen, 2001).

Moreover, the critical fact is that the severe pollution of the rivers in China has not yet been controlled. The overall environmental quality is deteriorating. The annual economic loss caused by water pollution occupies almost 1.5–3% of the GNP of the entire year (Qian et al., 2001). In 2002, only 29.1% of the 741 monitored sections of the seven major river basins met national water quality standards for Grades I–III, and 30% of the monitored sections were categorized as Grade IV or V water quality; the rest of them, 40.9%, were categorized as lower than Grade V (State Environmental Protection Administration of China, 2003). The major lakes in China are seriously polluted, and face critical eutrophication.

Furthermore, some incidental pollution by hazardous chemicals also has polluted waterbodies. For example, the chemical plant blast caused by violations of working rules discharged poisonous substances into the Songhua River in November, 2005, which forced the cities along the river, including Harbin, the capital of Heilongjiang Province with more than three million people, to temporarily suspend the water supply.

2.2 Example 1 – Drinking-water treatment

2.2.1 Demonstration plant for treatment of contaminated water with preoxidation/adsorption and biological filtration

Most of the water treatment plants in China employ conventional water treatment processes. However, the deterioration of the water resources has necessitated the upgrading of the conventional water treatment processes to guarantee the water quality. Various novel approaches, such as preoxidation and biological filtration, have been developed to upgrade the conventional water treatment process. An example of this is the Nanmen water treatment plant located in Jiaxing City, Zhejiang Province. This water treatment plant takes raw water from the great Jing-hang canal. With the increase in the amount of wastewater discharged into this canal and backward flow of wastewater from downriver because of the tide, the water in the Jing-hang canal was polluted. Consequently, the quality of the finished water from the conventional water treatment process could not comply with the standard.

Therefore, the original water treatment process had to be modified to guarantee the quality of the finished water. In the modified process, pre-chlorination was replaced by pre-oxidation with a chemical containing permanganate, and the rapid filtration tank was changed to a biological activated carbon filter (biofiltration). In addition, pre-ozonation and adsorption with powdered activated carbon (PAC) were added to modify the process. Biofiltration was introduced to the process to enhance the removal of organic compounds and ammonia nitrogen. Pre-ozonation and PAC absorption can improve the removal of organic matter effectively. In addition, pre-ozonation can increase the dissolved oxygen content in water, which benefits the following biofiltration. The dosages of ozone and PAC are 1.0 mg/L and 20 mg/L, respectively. The flow chart of the modified water treatment process is shown in Figure 4.

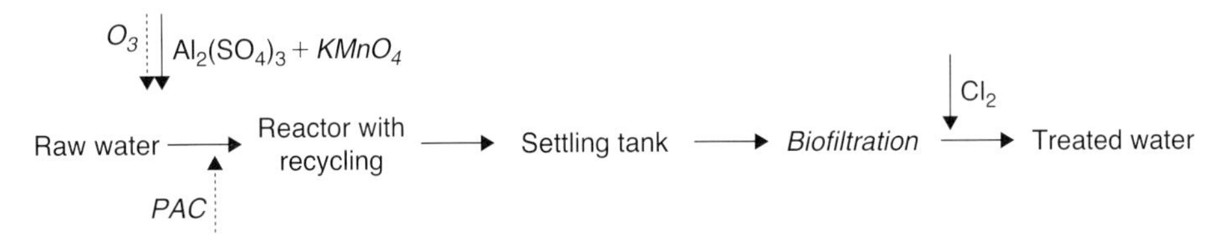

Figure 4 Flow chart of the modified conventional water treatment process at Nanmen plant (The dotted arrows and the *Italic* words represent the methods employed to upgrade the original process)

Table 1 Comparison of raw water and treated water quality between 2003 and 2004 at Nanmen plant

Parameter	*Standard*	*Raw water (X ± C)*[1]		*Treated water (X ± C)*[1]	
		2003	*2004*	*2003*	*2004*
Turbidity (NTU)	<1	107.3 ± 7.4	114.7 ± 8.2	0.72	0.7
Colour (degree)	<15	37.4 ± 0.7	36.4 ± 0.4	<15	<15
COD_{Mn} (mg/L)	<3	6.96 ± 0.14	7.49 ± 0.20	4.40 ± 0.22	3.71 ± 0.15
NH_4-N (mg/L)	<0.5[α]	2.14 ± 0.39	2.33 ± 0.32	1.03 ± 0.32	0.47 ± 0.17
Fe (mg/L)	<0.3	2.41 ± 0.12	3.51 ± 0.21	0.10 ± 0.02	<0.05
NO_2-N (mg/L)		0.299 ± 0.048	0.356 ± 0.024	0.017 ± 0.007	0.008 ± 0.006
Temperature (°C)		20~34	22~30		

[1] X is average value and C is standard error

The characteristics of the effluent from the conventional treatment process during June and July of 2003 and that from the modified treatment process during June and July of 2004 are summarized in Table 1 for comparison. In addition, the quality of raw water during different periods is also listed in Table 1 for reference.

Obviously, the quality of the effluent considerably improved after the treatment process had been modified. The removal efficiencies of COD_{Mn} and ammonia nitrogen increased from 36.8% to 50.5% and from 51.9% to 79.8%, respectively. However, the COD_{Mn} of the treated water from the modified treatment process cannot meet with the drinking-water standards. Further research should be carried out to improve the removal of organic matter from the raw water.

2.2.2 *Demonstration plant – Application of ozonation and catalytic ozonation for upgrading a drinking-water treatment plant*

As many water resources in China are contaminated with micropollutants and the drinking-water standard is becoming more stringent, China is paying increasing attention to advanced water treatment processes. Among the advanced water treatment processes, ozonation or catalytic ozonation followed by biological activated carbon filtration has proven to be an effective process for removing micropollutants from water. Catalytic ozonation aims to degrade persistent organic pollutants and endocrine disrupting substances in water. The biological activated carbon filter can further degrade the intermediate organic micropollutants, and reduce the assimible organic carbon (AOC).

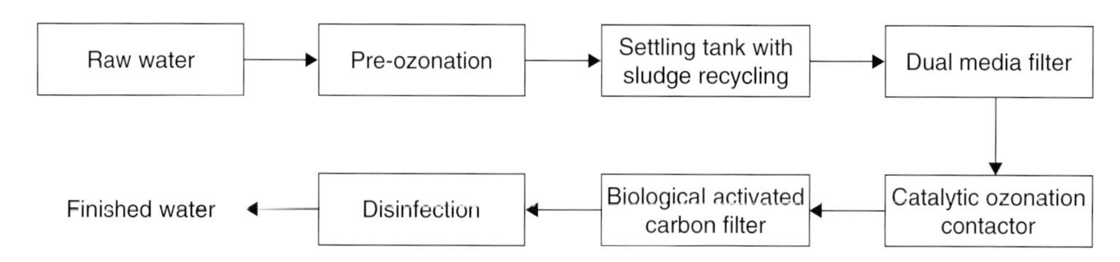

Figure 5 The flow chart of the new water treatment process

A typical advanced water treatment process was adopted at the Shijiuyang Water Treatment Plant (Shijiuyang WTP), which accounts for 80% of the water supply in Jiaxing city. The water source of Shijiuyang WTP is a branch of the great Jing-hang channel. With the development of industry and agriculture on both sides of the channel, substantial amounts of pollutants are discharged into this channel, leading to the deterioration of the water source quality of the Shijiuyang WTP. In the past, a conventional water treatment process consisting of coagulation, flocculation, sedimentation, filtration and chlorination units was employed in Shijiuyang WTP. With this treatment process, the effluent of Shijiuyang WTP could barely meet the national water quality standard, especially the organic matter content in the treated effluent. To guarantee the water quality, between 2003 and 2004, Shijiuyang WTP upgraded the existing treatment process by installing an intermediate ozonation tank and biological activated carbon (BAC) filter (single ozonation – BAC) before the chlorination tank (this process is referred to the conventional ozone process in the following parts).

Furthermore, to satisfy the increased water demand, the plant constructed a new water treatment process composed of both conventional and advanced treatments, as presented in Figure 5. In the new process, catalytic ozonation – biological activated carbon filter (catalytic ozonation – BAC) was selected as the advanced treatment unit. The difference between catalytic ozonation and single ozonation is that the former unit is filled with efficient solid catalysts which can generate hydroxyl radicals.

Long-term experiments indicate that catalytic ozonation technology is capable of removing complex organic pollutants, trace pollutants, DBPs precursors and mutagen. As a result, the use of the catalyst decreased the finished water's mutagenicity from 25% to 0. The full-scale operation of the Shijiuyang WTP Plant demonstrated that the catalytic ozonation process can reduce the formation potential of bromate and AOC by 51.7% and 58.85%, respectively. It was also noted that the dosage of ozone in the catalytic ozonation process is only 60% of that applied in the ozonation process to reduce the COD_{Mn} concentration in the influent from 2.20–5.80 mg/L to 3.0 mg/L. To reduce the COD_{Mn} concentration in the effluent to 2.0 mg/L, the introduction of the catalyst can save 25–30% of the ozone.

Furthermore, catalytic ozonation is effective in reducing the load of organic pollutants, and improving the biological activity and the biodegradability of organic substances for following BAC treatment. Of the organic pollutants removed in the whole advanced treatment process, 30–94.3% was removed in the catalytic ozonation contactor. The bio-activity in the following process was enhanced by 35–60% and the content of BDOC in the effluent of the ozonation contactor was increased by 110% due to the

Table 2 Quality of raw water and finished water from different processes at Shijiuyang water treatment plant

Parameters	*Raw water*	*Finished water from the conventional oxidation process*	*Finished water from the catalytic ozonation process*
Colour /SU	42	<5	<5
Turbidity /NTU	12	0.25	0.2
Temperature/°C	32	30	30
pH	8.02	7.78	7.97
Total iron /mg · L^{-1}	0.22	0.08	0.05
Total manganese /mg · L^{-1}	0.356	<0.05	<0.05
Cuprum /mg · L^{-1}	<0.1	<0.1	<0.1
Zincum /mg · L^{-1}	<0.1	<0.1	0.299
Volatile phene / mg · L^{-1}	<0.002	<0.002	<0.002
Sulphate /mg · L^{-1}	63.7	57.8	52.3
Chloride /mg · L^{-1}	51.9	54.1	53.7
Dissolve solid /mg · L^{-1}	365	341	309
Fluoride /mg · L^{-1}	0.69	0.69	0.63
Cyanide /mg · L^{-1}	<0.002	0.002	<0.002
Arsenic /mg · L^{-1}	0.007	0.004	0.003
Selenium /mg · L^{-1}	<0.002	<0.002	<0.002
Hydrargyrum /mg · L^{-1}	0.0002	<0.0001	<0.0001
Chromium /mg · L^{-1}	0.006	<0.004	<0.004
NO_3-N mg · L^{-1}	<0.04	1.03	1.57
Bacteria count/CFU/mL	68000	15	1
Residual chlorine /mg · L^{-1}	—	0.5	0.4
Gross α radioactivity/Bq · L^{-1}	<0.016	<0.016	<0.016
Gross β radioactivity/Bq · L^{-1}	0.302	0.161	0.223
Clofenotane /μg · L^{-1}	<1	<1	<1
Chloroform /μg · L^{-1}	2.2	3	3.4
Carbon tetrachloride /μg · L^{-1}	<0.3	<0.3	<0.3
NH_3-N/mg · L^{-1}	0.77	0.32	0.18
NO_2-N/mg · L^{-1}	0.076	<0.001	<0.001
COD_{Mn}/mg · L^{-1}	10.9	3.04	2.22
Alkalinity/mg · L^{-1}	119.1	94.5	85.3
Aluminum/mg · L^{-1}	0.16	0.056	0.043
TOC/mg · L^{-1}	10.69	3.21	1.89
Formaldehyde/mg · L^{-1}	<0.05	0.15	0.11
BrO_3^-/μg · L^{-1}	<2	8.15	<2

introduction of the catalyst. Compared to ozonation, catalytic ozonation is better at removing algae, chlorophyll a and enhancing the nitrification of NH_4-N. In summary, the capacity of organic matter removal with catalytic ozonation was increased by 110–143% and the lifetime of the BAC was increased by about 8 months because of the application of the catalyst. Table 2 summarizes the qualities of the water source, finished water from the control process and that from the new catalytic ozonation process.

2.3 Example 2 – Municipal wastewater treatment

With the rapid urbanization and huge production of sewage, municipal wastewater treatment has drawn increased attention in China since the 1980s. Generally, the conventional activated sludge process or the modified activated sludge process was adopted by the municipal wastewater treatment plant (WTP) constructed in the 1980s. In the 1990s, the A/O process and A^2/O process became dominant in the newly constructed municipal WTP, because nitrogen and phosphorus pollution had to be controlled. Later on, oxidation ditch and SBR processes became increasingly prevalent. Here, two large WTPs with a conventional activated sludge process or a modified activated sludge process are introduced.

2.3.1 *Demonstration plant – Gaobeidian wastewater treatment plant*

The Gaobeidian wastewater treatment plant (WTP) is the first large-scale municipal wastewater treatment plant constructed in Beijing and its treatment capacity is 1 million m^3/day. The treatment process adopted in this WTP is as follows:

The quality of the effluent from this process is very stable and can meet the Grade I of the national wastewater discharge standard, with BOD_5 ranging from 4.7 to 7.8 mg/L, COD ranging from 20.3 to 44.6 mg/L and SS varying from 14 to 23 mg/L. Part of the effluent (10,000 m^3/day) from this process is further treated with coagulation, settlement and sand filtration so that it can be reused.

2.3.2 *Demonstration plant – Dongjiao wastewater treatment plant*

Dongjiao wastewater treatment plant (WTP), with a treatment capacity of 400,000 m^3/day, is located in Tianjin, one of the municipalities in China. Dongjiao WTP adopted the conventional activated sludge treatment process, as demonstrated in Figure 7.

The quality of the effluent in summer is better than that in winter, with the average BOD_5 in the range 8.6–23.1 mg/L, COD in the range 72–110 mg/L and SS in the range 16–65 mg/L.

2.4 Example 3 – Advanced treatment and water reuse

The water shortage problem is very serious in China's cities due to environmental pollution and the scarcity of water resources. Water regeneration with the effluent from secondary wastewater treatment plants by advanced treatment can mitigate the contradiction between water supply and water demand. As the effluent from secondary wastewater treatment plants is easy to collect and can undergo centralized treatment, it is more practical than seawater or rainwater as the secondary water resource for cities, after advanced treatment. In China, regenerated water is generally used for agriculture, industry, multi-purpose municipal usage, re-irrigation, etc., depending on its quality. Membrane separation, advanced oxidation, ion exchange, adsorption, filtration and disinfection, etc. are the technologies frequently employed for advanced treatment of effluent from the secondary treatment process, e.g., the two typical water reuse demonstration plants in China that are introduced below.

2.4.1 *Demonstration plant – Water reuse in Dalian Chunliuhe wastewater treatment plant*

Chunliuhe secondary WTP is the first municipal wastewater treatment plant in Dalian, with a treatment capacity of 60,000 m^3/day. In 1992, the national demonstration water reuse project of 10,000 m^3/day was carried out in Chunliuhe WTP and the finished water was reused for garden irrigation and industrial cooling and recycling.

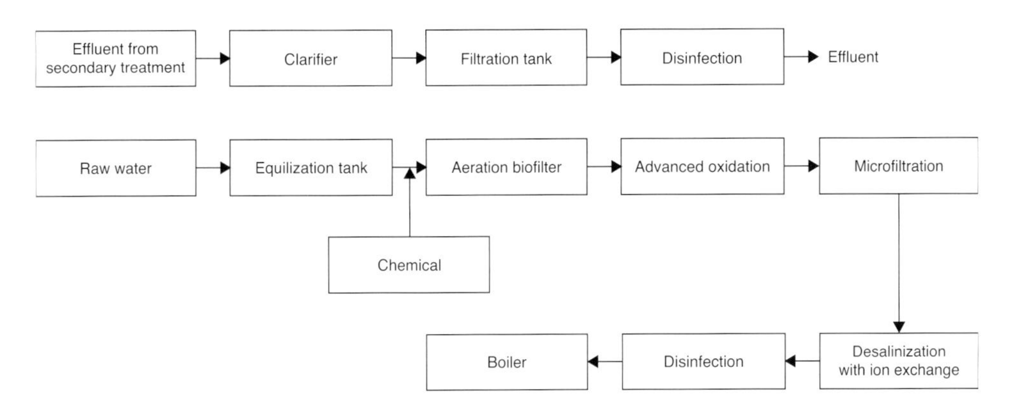

Figure 6 Treatment process for water regeneration in Chunliuhe WTP

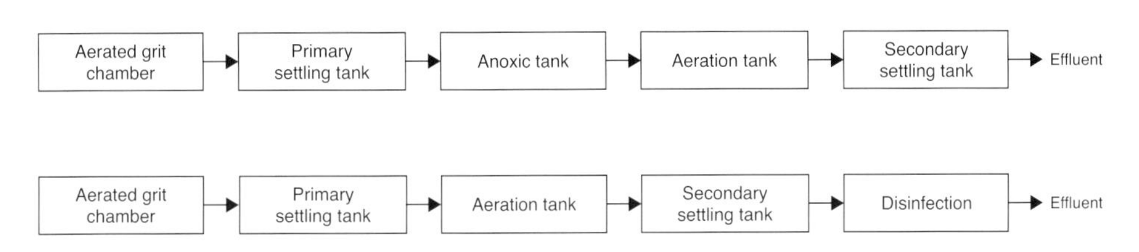

Figure 7 Treatment process for water regeneration at Pingfang area of Harbin

One-sixth (10,000 m^3/day) of the effluent from the secondary treatment was further treated with the following process, as shown schematically in Figure 6, to meet the standard for multi-purpose domestic water uses. The characteristics of the finished water from this process are as follows: BOD $\leqslant$ 6 mg/L, COD $\leqslant$ 15 mg/L and SS $\leqslant$ 15 mg/L.

2.4.2 Demonstration plant – Water reuse engineering at Pingfang area of Harbin

To reduce the discharge of wastewater, a wastewater reclamation plant, which treats the municipal wastewater and part of the treated industrial wastewater for water reuse, was built in the Pingfang area of Harbin. The process adopted is illustrated in Figure 7. The effluent from this process can be reused as boiler supplement water.

3 SUMMARY

Water is of vital importance to human societies and environments. How to achieve sustainable development on the basis of sustainable use of water is of great concern to China. Over the last five decades, water treatment has attracted great attention in China and achieved rapid progress. China has constructed 277,000 kilometres of river dikes and 85,000 reservoirs. Flood control systems in major river basins have been primarily established, which helped to defeat major floods over the last 50 years. The total annual water supply capacity has increased to 600 billion m^3, guaranteeing water use for 56 million hectares of irrigated farmland, over 600 cities and a variety of industrial sectors. The urban wastewater treatment percentage has reached 45%. The great achievements in water development mean that China can now support 20% of the world population with only 6.8% of the world's cultivated land and 6% of the world's water resources.

However, unfavourable water conditions, rapid economic development and a large population have imposed heavy pressure on water resources, which can be reflected with the following aspects. Water shortage causes further conflict between water demand and supply. Flood control capacity is still not enough and flood damages are serious. Water quality has deteriorated, which further intensifies the water shortage situation. Although China has accumulated some experiences in urban water management, the present systems in China are not sustainable for the future. To solve the water-related problems, it is recommended: 1) to develop a water-saving society and improve water use efficiency; 2) to highlight the priority of water environment protection in the line of the recycling economy; 3) to construct water allocation projects to promote regulation on uneven natural water distribution; and 4) to rehabilitate and protect ecology and the environment in China.

REFERENCES

Dokulil, M., Chen, W. and Cai, Q. 2000. Anthropogenic Impacts to Large Lakes in China: The Tai Hu Example, *Aquatic Ecosystem Health and Management*, Vol. 3, No. 1, pp. 81–94.

Feng, S., Li, L.X., Duan, Z.G. and Zhang, J.L. 2007. Assessing the Impacts of South-to-North Water Transfer Project with Decision Support Systems, *Decision Support Systems*, Vol. 42, No. 4, pp. 1989–2003.

He, P.J., Phan, L., Gu, G.W. and Hervouet, G. 2001. Reclaimed Municipal Wastewater – A Potential Water Resource in China, *Water Science Technology*, Vol. 43, pp. 51–58.

He, R.J. 2006. The Present Situation and Prospect of Urban Sewage Treatment (in Chinese), *Sci-Tech Information Development & Economy*, Vol. 16, pp. 181–182.

Huang, Z.J. 1998. Problem and Countermeasures of Urban Water Supply, *Water and Wastewater Engineering* (in Chinese), Vol. 24, pp. 18–20.

Kasprzyk-Hordern, B., Ziółek, M. and Nawrocki, J. 2003. Catalytic Ozonation and Methods of Enhancing Molecular Ozone Reactions in Water Treatment, *Applied Catalysis B: Environmental*, Vol. 46, pp. 639–669.

Langlais B., Reckhow D.A. and Brink D.R. 1991. *Ozone in Water Treatment: Application and Engineering*. Lewis Publishers.

Ma, J. and Graham, N. 1996. Controlling the Formation of Chloroform by Permanganate Preoxidation—Destruction of Precursors, *Journal of Water SRT—Aqua*, Vol. 45, No. 6, pp. 308–315.

Ma, J. and Graham, N. 1997. Preliminary Investigation of Manganese-catalyzed Ozonation for the Destruction of Atrazine, *Ozone Science and Engineering*, Vol. 19, pp. 227–240.

Ma, J. and Graham, N. 1999. Manganese-catalyzed Ozonation for the Destruction of Atrazine – Effect of Humic Substances, *Water Research*, Vol. 33, pp. 785–793.

Ma, J. and Graham, N. 2000. Degradation of Atrazine by Manganese-catalyzed Ozonation – Influence of Radical Scavengers, *Water Research*, Vol. 34, pp. 3822–3828.

Ma, J. and Li, G.B. 1993. Laboratory and Full-scale Plant Studies of Permanganate Oxidation as an Aid to the Coagulation, *Water Science and Technology*, Vol. 27, No. 11, pp. 47–54.

Ma, J. and Sui, M.H. 2004. Degradation of Refractory Organic Pollutants by Catalytic Ozonation – Activated Carbon and Mn-loaded Activated Carbon as Catalysts, *Ozone Science and Engineering*, Vol. 26, pp. 3–10.

Ma, J. and Sui, M.H. 2005. Effect of pH on MnOx/GAC Catalyzed Ozonation for Degradation of Nitrobenzene, *Water Research*, Vol. 39, No. 3, pp. 779–786.

Ma, J., Graham, N. and Li, G. 1997. Effect of Permanganate Preoxidation in Enhancing the Coagulation of Surface Waters – Laboratory Case Studies, *Journal of Water SRT—Aqua*, Vol. 46, No. 1, pp. 1–10.

Ma, J., Li, G.B. and Graham, N. 1994. Efficiency and Mechanism of Acrylamide Removal by Permanganate Oxidation, *Journal of Water SRT—Aqua*, Vol. 43, No. 6, pp. 278–295.

Ma, J., Li, G.B., Chen, Z.L., Xu, G.R. and Cai, G.Q. 2001. Enhanced Coagulation of Surface Waters with High Organic Content by Permanganate Preoxidation, *Water Science and Technology: Water Supply*, Vol. 1, No. 1, pp. 51–61.

Qian, Z., Zhang, G. et al. 2001. *Report of Strategic Study on Water Resources for Sustainable Development of China* [M]. China Water Power Press, Beijing, pp. 28–31, 1–3 [in Chinese].

Shen, L., Lin, G.F., Tan, J.W. and Shen, J.H. 2000. Genotoxicity of Surface Water Samples from Meiliang Bay, Taihu Lake, Eastern China, *Chemosphere*, Vol. 41, No. 1–2, pp. 129–132.

State Environmental Protection Administration of China, Environmental status bulletin of China (2002), 2003.

Tselentis, Y. (1996). Effluent Reuse Options in Athens Metropolitan Area: A Case Study, *Water Science Technology*, Vol. 33, pp. 127–138.

Varis, O. and Vakkilainen, P. 2001. China's 8 Challenges to Water Resources Management in the First Quarter of the 21st Century, *Geomorphology*, Vol. 41, pp. 93–104.

Wang, T. and Lou, S.Y. 2004. Investigation on Urban Wastewater Treatment Processes in China and Technical and Economic Evaluation [in Chinese]. *Water & Wastewater Engineering*, Vol. 30, pp. 1–4.

Xu, M.Q., Cao, H., Xie, P., Deng, D.G., Feng, W.S. and Xu, J. 2005. The Temporal and Spatial Distribution, Composition and Abundance of Protozoa in Chaohu Lake, China: Relationship with Eutrophication, *European Journal of Protistology*, Vol. 41, No. 3, pp. 183–192.

Yang, S.Y. 1997. Problems and Countermeasures of China's Water Resources, *Environmental Protection* [in Chinese], Vol. 11, pp. 6–8.

Case Study V

Challenges for urban water management in Cairo, Egypt: the need for sustainable solutions

El Said M. Ahmed[1] and Mohamed A. Ashour[2]

[1]Arizona Department of Water Resources, Phoenix, Arizona, USA
[2]Assiut University, Assiut, Egypt

1 EGYPTIAN WATER SYSTEM

Egypt lies in the north-eastern corner of the African continent and has a total area of about 1 million km^2. The majority (about 97%) of the country area is desert land. The Egyptian terrain consists of a vast desert plateau interrupted by the Nile Valley and Delta which occupy about 4% of the total country area. Most of the cultivated land is located close to the banks of the Nile River, its main branches and canals, and in the Nile Delta. The total cultivated area is about 3% of the total area of the country. Hot dry summers and mild winters characterize Egypt's climate. Rainfall is very low, irregular and unpredictable. Annual rainfall ranges between a maximum of about 200 mm in the northern coastal region to a minimum of nearly zero in the south, with an annual average of 51 mm. Summer temperatures are extremely high, reaching 38°C to 43°C, with extremes of 49°C in the southern and western deserts. The northern areas on the Mediterranean coast are much cooler, with 32°C as a maximum.

The population is estimated at 73.4 million (2004), with an average annual growth rate of 1.8%. The rural population is 58% of the total population. The overall population density is 73 inhabitants/km^2; however, with about 97% of all people living in the Nile Valley and Delta, population density reaches more than 1,165 inhabitants/km^2 in these areas, while in the desert it drops to only 1.2 inhabitants/km^2. In 2000, about 96% of the rural population and 99% of the urban population had access to improved drinking-water sources, with an average of 97% of the total population. Almost 100% of the urban population and 96% of the rural population had access to improved sanitation, with an average of 98% of the total population.

2 WATER RESOURCES

The Egyptian territory comprises the following river basins:

- The Northern Interior Basin, covering 520,881 km^2 or 52% of the total area of the country in the east and south-east of the country. A sub-basin of the Northern Interior Basin is the Qattara Depression.

- The Nile Basin, covering 326,751 km^2 (33%) in the central part of the country in the form of a broad north–south strip.
- The Mediterranean Coast Basin, covering 65,568 km^2 (6%).
- The North-east Coast Basin, a narrow strip of 88,250 km^2 along the coast of the Red Sea (8%).

The River Nile is the main source of water for Egypt, with an annual allocated flow of 55.5 km^3/yr under the Nile Waters Agreement of 1959. Internal surface water resources are estimated at 0.5 km^3/yr. This brings the total actual surface water resources to 56 km^3/year. The Nubian Sandstone aquifer located under the Western Desert is considered an important groundwater source. The volume of groundwater entering the country from Libya is estimated at 1 km^3/yr. Internal renewable groundwater resources are estimated at 1.3 km^3/yr, bringing the total renewable groundwater resources to 2.3 km^3/yr. The main source of internal recharge is percolation from irrigation water in the Valley and the Delta. The total actual renewable water resources of the country is thus 58.3 km^3/yr.

All drainage water in Upper Egypt, south of Cairo, flows back into the Nile and the irrigation canals; this amount is estimated at 4 km^3/yr. Drainage water in the Nile Delta is estimated at 14 km^3/yr. Treated domestic wastewater in 2001/02 was estimated at 2.97 km^3/yr. There are several desalination plants on the coasts of the Red Sea and the Mediterranean to provide water for seaside resorts and hotels; the total production in 2002 was estimated at 100 million m^3. Estimates of the potential of non-renewable groundwater in the eastern and western deserts, mainly from the Nubian Sandstone aquifer, vary from 3.8 km^3/yr to 0.6 km^3/yr; the latter estimate is defined as an indicator of exploitability over a period of time, where the time is not given.

3 WATER USE

Total water withdrawal in 2000 was estimated at 68.3 km^3. Figure 1 shows the total water withdrawal by sector in Egypt. This included 59 km^3 for agriculture (86%), 5.3 km^3 for domestic use (8%) and 4.0 km^3 for industry (6%). Apart from that, 4.0 km^3 were used for navigation and hydropower.

Groundwater extraction in 2000 was 7.043 km^3 comprising:

- 6.127 km^3 from the Nile Basin (seepage waters)
- 0.825 km^3 from the eastern and western deserts, i.e. mainly the Nubian Sandstone aquifer
- 0.091 km^3 from shallow wells in Sinai and on the north-western coast.

Reuse of agricultural drainage water, returned to the rivers, in irrigation amounted to 4.84 km^3/yr in 2001/02. Of the 2.97 km^3/yr of treated wastewater, 1.5 km^3/yr is reused for irrigation, while the rest is pumped into main drains where it mixes with drainage water and is then used for irrigation. Treated wastewater is usually used for landscape irrigation of trees in urban areas and along roads.

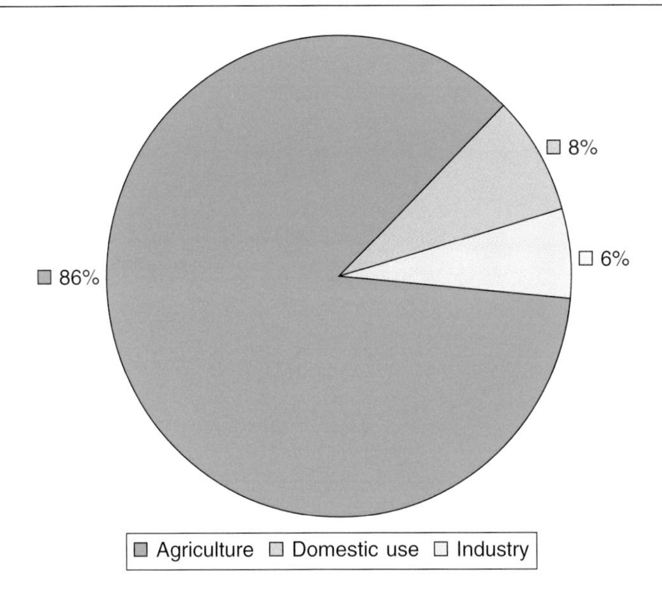

Figure 1 Total water withdrawal by sector, Egypt, 2000

Under the 1959 Nile Waters Agreement between Egypt and Sudan, Egypt's share of the Nile flow is 55.5 km^3/yr. The agreement was based on the average flow of the Nile during the period 1900–1959, which was 84 km^3/yr at Aswan. Average annual evaporation and other losses from the Aswan High Dam and reservoir (Lake Nasser) were estimated at 10 km^3/yr, leaving a net usable flow of 74 km^3/yr, of which 18.5 km^3/yr was allocated to Sudan and 55.5 km^3/yr to Egypt. If conditions permit the completion of the development projects on the Upper Nile, Egypt's share in the Nile water will increase by 9 km^3. This amount includes 1.9 km^3 and 1.6 km^3 respectively from the first and second phases of the Jonglei canal project in southern Sudan. Two other projects in the upstream swamps are expected to provide 5.5 km^3.

4 WATER SUPPLY SYSTEM AND SEWERAGE SYSTEM IN CAIRO

Cairo is the capital of Egypt and is one of the largest cities in Africa. Located on both banks of the Nile River (Figure 2) near the head of the river's delta in northern Egypt (Figure 3). The site has been settled for more than 6,000 years and has served as the capital of numerous Egyptian civilizations. Greater Cairo is spread across three of Egypt's administrative governorates: the east bank portion is located in Al Qalyobīyah Governorate, while the west bank is part of the governorates of Al Jīzah and Al Qalyobīyah. Cairo is marked by the traditions and influences of the East and the West, the ancient and the modern.

The city of Cairo covers an area of more than 453 sq km (more than 175 sq mi), though it is difficult to separate the city from some of its immediate suburbs. Bracketed by the desert to the east, south and west, and bounded by the fertile Nile Delta to the

Figure 2 Cairo located on both banks of the Nile River (See also colour plate 28)

north, Cairo spreads farther on the east bank than the west. Cairo also includes several river islands, which play an important role in the life of the city. As the region's principal commercial, administrative and tourist centre, Cairo contains many cultural institutions, business establishments, governmental offices, universities and hotels, which together create a dense pattern of constant activity affecting its water and sewerage systems. Cairo's drinking water is drawn from the Nile. In general, in the 1970s, the river's water quality appeared quite good, and that continued into the 1980s, but following that decade deterioration indicators started to appear which monitoring over the entire Nile confirmed. This deterioration was due to increased industrial and agricultural discharges, and also (moderate) contamination from human sewage.

The Cairo Water Authority has 13 clean water treatment plants. Table 1 summarizes the clean water treatment plants, their original design capacities, expansions, current capacities, and the served areas. Also Figure 4 shows the approximate locations and distribution of the clean water treatment plan in the greater Cairo area. The water supply is drawn from intake points in the middle of the Nile. The finished water goes to storage or pump stations for distribution; at this point, as it enters the distribution system, Cairo's drinking water is nearly always clean. However, some problems in the water distribution system or storage sometimes lead to erratic water supplies and/or contamination entering the drinking water in several areas. Erratic water pressure and unreliable supply may cause pollution from contaminated groundwater or sewage from leaking drains and sewers entering the drinking-water distribution system through damaged joints.

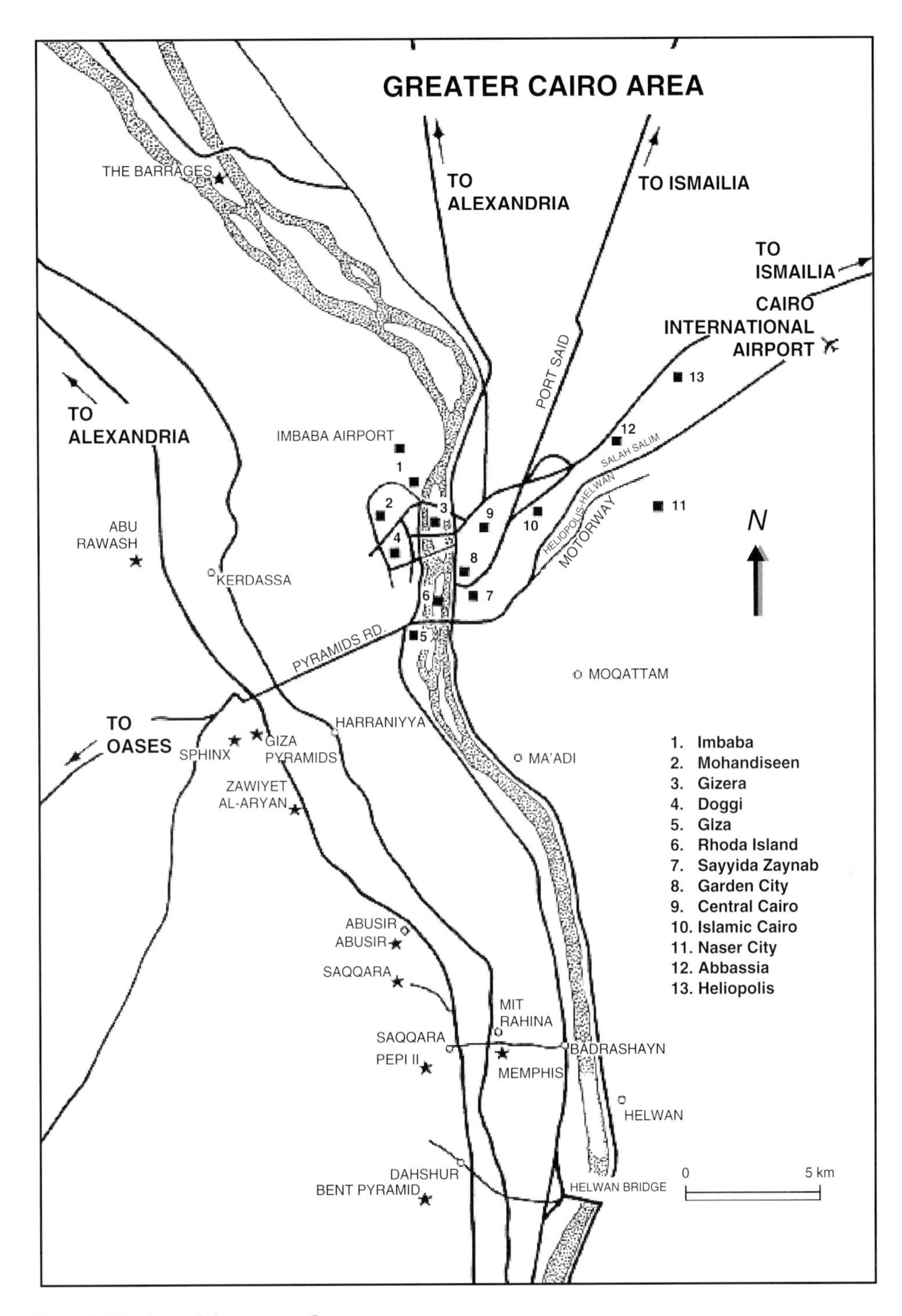

Figure 3 Districts of the greater Cairo area

Table 1 Clear water treatment plants in the greater Cairo area

Clear water treatment plant	Year built	Original capacity (1000 m^3/day)	Expansions	Today's capacity (1000 m^3/day)	Served areas
Msstrud	1977	500	Additional 600,000 m^3/day – 1994	1100	Misr El-Gededa, Nasr City, Almazah, El-Zaytoon, Ain Shamss, El-Mataria, El-Marg, and Madenat El-Salam
Amereha	1962	300	Additional 150,000 m^3/day – 1996	450	El-Sharapya, Gamrah, El-Zaytoon, Misr El-Gededa, Nasr City
Rood El-Faragg	1903	Unknown	On 1916 and 1964 it was upgraded to 100,000 m3/day. Additional 800,000 m^3/day – 1997	900	Rood El-Faragg, Shoubra, El-Saheel, Weest El-Madenah, El-Darasah, El-Gamaliah, and El-Sayda Zynab
Fostat	1988	450	Additional 300,000 m^3/day – 1994 and 300,000 m^3/day – 2000	1050	Dar El-Salam, Misr El-Kadema, El-Kalaa, El-Basateen, El-Maady, El-Katamia, Nasr city, and El-Maasarah
Maady	1900	90	–	90	Maady, Maady El-Sarayat, Torah El-Balad, and Torah El-Assmnt
Teben	1973	150	Under construction to add 200,000 m^3/day	150	Teben, Atlass Residential areas, 15 May, and Steel factory in Helwan.
Roudah	1963	180	–	180	Manial El-Roudah, Missr El-Kadema, El-Madapeg, Abo-Elsood, and Ein El-Sierah
Kafr El-Alow	1921	110	–	110	Kafr El-Alow and Helwan El-Balad
North Helwan	1980	170	Additional 200,000 m^3/day – 2005	370	Helwan and Maasarah
Dahab Island	1976	220	Additional 50,000 m^3/day – 1996 and 200,000 m^3/day – 1999	470	Dahab Island, South Giza, Moneeb, and Elomranya
Embaba	1986	300	Additional 200,000 m^3/day – 1997 and 200,000 m^3/day – 1999	700	Elwaraak, Embaba, dokki, Mohandseen, Giza, Bolak Eldakror, Faysal, and Elharam
Giza	1985	150	–	–	Giza City and Been Elsarayat
Shoubra El-Keema	1966	200	Under construction to add 200,000 m^3/day	200	Shoubra El-Keema town and east and west sides of Shoubra El-Keema town

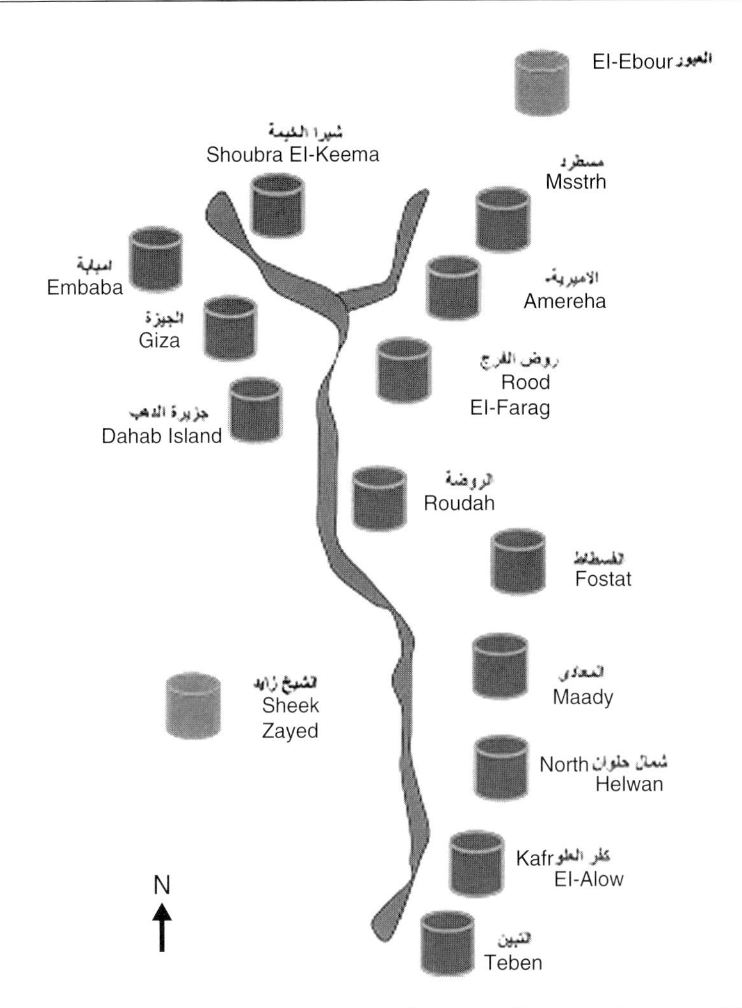

Figure 4 Clear water treatment plants in the greater Cairo area (See also colour plate 29)

Cairo's first formal sewerage system was built by the British just before World War I for a city of about one million people. It was grossly inadequate. The city over the last two decades has built an entirely new sewerage system, including an enormous treatment plant north of the city. Because the old system was so overburdened, this did not mean simply expanding the old system, but building a new one from scratch.

The new wastewater system consists of six wastewater treatment plants (WWTP). The Helwan WWTP is located south of Cairo near the east bank of the River Nile. The Berka WWTP, Balas WWTP and Gabal Asfar WWTP are located on the north-east side of Cairo. The Zenein WWTP and Abo Rawash WWTP are located on the west side of Cairo. Figure 5 shows the approximate location of Cairo's Wastewater Treatment Plants distributed around the greater Cairo area. Table 2 summarizes the design capacities of the plants.

The core of the new system was a 'trunk line' – the central line to which others attach (see Figure 5). This trunk line is 5 metres (16.4 feet) in diameter, and extends in a sloping, gravity-fed descent from south to north through the city until it ends up by

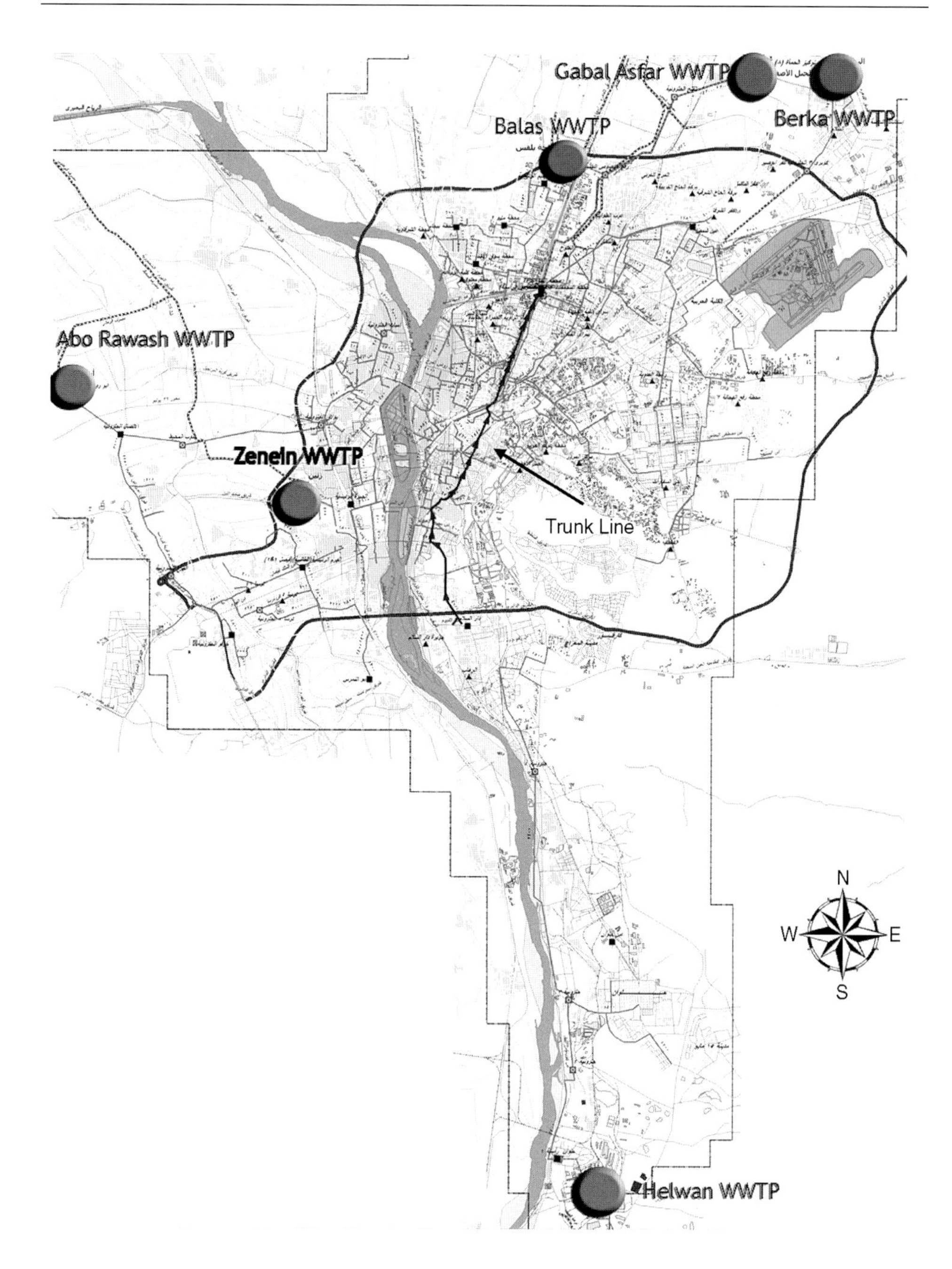

Figure 5 Wastewater treatment plants in the greater Cairo area (See also colour plate 30)

the new treatment plant at a depth of 25 metres. Its core phases are now completed, and it is one of the largest sewerage projects in the world. But the city is still struggling to connect everyone to it. Millions of the city's residents live in illegally built concrete and brick apartments that scatter out across the desert in endless waves.

Table 2 Water treatment plants in the greater Cairo area

Wastewater treatment plant	*Design capacity (1000 m^3/day)*
Helwan	350
Zenein	330
Abo Rawash	400
Berka	600
Balas	600
Gabal Asfar	–

5 MAJOR PROBLEMS WITH THE WATER AND SEWERAGE SYSTEMS IN CAIRO

Surface waters from the Nile River are the major source of the bulk water supply in Cairo. The city draws all of its water from the Nile, which is then purified and pumped into pipes for distribution all over the city. However, expansion of the city has put a large strain on the water supply, which has in some cases led to low water pressure, essentially rendering the supply useless. Many of the pipes that carry drinking water are either made of or contain a significant amount of lead.

The majority of the Cairo residents receive treated drinking water in individual connections in their homes, but it is not known with certainty exactly what percentage of people have this service. The Cairo Water Authority has cited an estimate of over 90%. However many buildings served may not have indoor plumbing that further distributes the water to apartments. Also, it has been noted that since the water service is often erratic in terms of pressure, hours of service and volume, interrupted service during peak hours is a common complaint (many residents must store water in the evening in bathtubs for use in the morning). Some of East Bank residents were served by public fountains, rather than individual household connections. People have to wait for a long time in queues at communal taps, or, where there is not even communal taps, the (poorer) people have to buy the water through unsafe containers at very high prices. Thus, individuals relying on a neighbourhood standpipe for water will individually use much less water than will people with connections in their home. Additionally, many of the pipes that carry drinking water run right next to the lines that carry raw sewage and leaks are a constant problem in the aging infrastructure.

One final factor that has led to significant problems involves the sewerage system: the daily water usage far exceeds the capacity of the sewerage system. This leads to standing pools of raw sewage in the streets and in the underground water table, as well as leaks into the Nile and other sources of clean water.

As a conclusion, Cairo's drinking water is quite well treated but wastewater treatment has many severe deficiencies. Conveying Cairo's wastewaters through agricultural drains is just shifting the environmental problem to other regions, so the original problem still remains.

6 SUSTAINABILITY OF THESE SYSTEMS FOR THE FUTURE

It is essential that new water sources are found, and new agricultural areas and cities outside the valley are created. The only direction is to expand to the dry lands, desert, which is described as 'Egypt's last frontier'. Such a solution could provide an alternative to Cairo and relieve the pressure coming from a growing population. There are already several new cities and agricultural areas in the desert.

Besides finding new sources, water conservation is another strategy in water management. In the case of Cairo this needs the commitment of government institutions and international donors, as well as Cairo's residents. If water consumption continues to grow intensively, Egypt will have to rely on extreme measures: use the non-renewable groundwater aquifers and expensive desalinization of seawater.

For a city with more than 16 million people and almost no open space, and to keep sprawl from gobbling up agricultural land up and down the Nile, the government is redirecting growth to the east and west, into the desert. This has meant the establishment of several 'new cities'. One of them, New Cairo, has a growth area of 43,000 acres – the equivalent of nine Manhattans in land area – and 2 to 3 million people are expected to live there.

As a final conclusion, Cairo's water and wastewater systems could be sustainable if the population growth controlled the redirection growth to the east and west into the desert succeeds.

REFERENCES

Drainage Research Institute. 1989. *Land Drainage in Egypt.* Amer & de Ridder Editions.

Engelman, R. and Le Roy, P. 1993. Sustaining Water, Population and the Future of Renewable Water Supplies. *Population Action International, Population and Environment Program,* Washington, D. C., pp. 302–318.

Ground Water Research Institute. 2001. *The Groundwater Sector Plan, National Level.* National Water Research Center. Cairo, Egypt.

International Fund for Agricultural Development (IFAD). 2002. *Arab Republic of Egypt. Country Strategic Opportunities Paper (COSOP).* Rome.

Ministry of Agriculture and Land Reclamation. 1993. *Agricultural Statistics Data, 1989/1990, Country total.* General Administration of Agricultural Census, Sector of Economic Affairs. Arab Republic of Egypt.

Ministry of Agriculture and Land Reclamation. 2002. *Agricultural Statistics, Volume 1, Winter Crops 2002.* Sector of Economic Affairs. Arab Republic of Egypt.

Ministry of Agriculture and Land Reclamation. 2003. *Agricultural Statistics, Volume 2, Summer and Nili Crops 2002.* Sector of Economic Affairs. Arab Republic of Egypt.

Ministry of Agriculture and Land Reclamation. 2003. *Agricultural Statistics Data, 1999/2000.* General Administration of Agricultural Census, Sector of Economic Affairs. Arab Republic of Egypt.

Ministry of Planning. 2002. *The Fifth Five-Year Plan for Economic and Social Development 2002–2007.* Cairo, Egypt.

Ministry of Water Resources and Irrigation (ed.). 2002. *Adopted Measures to Face Major Challenges in the Egyptian Water Sector.* World Water Council.

National Planning Institute. 2001. A New Approach for Optimum Water Use in Egypt with Emphasis on Irrigation Water. *Series of Planning and Development Issues No.139.* Cairo, Egypt [in Arabic].

Parker, D. S. 1987. Wastewater Technology Innovation for the Year 2000. *National Conference on Environmental Engineering,* Orlando, Fl., USA.

Index

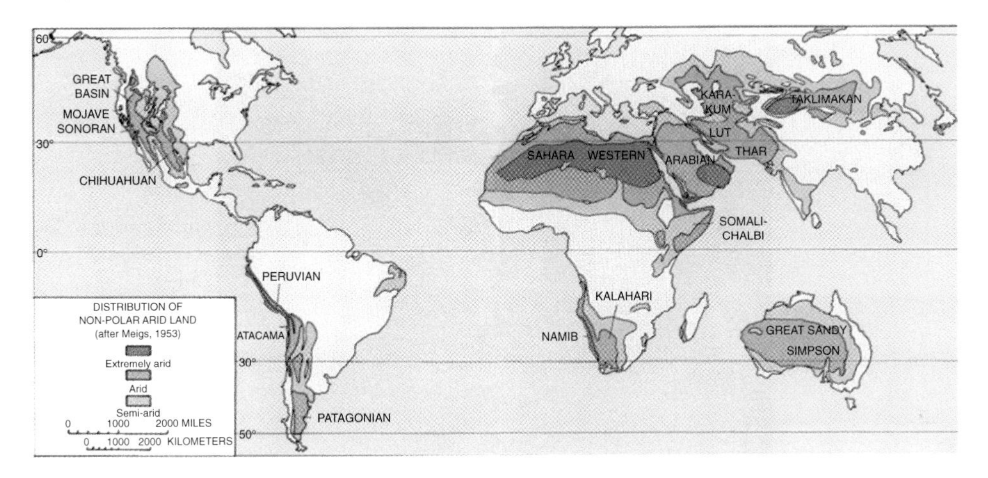

Plate 1 Distribution of non-polar arid land (after Meigs, 1953)

Source: http://pubs.usgs.gov/gip/deserts/what/world.html

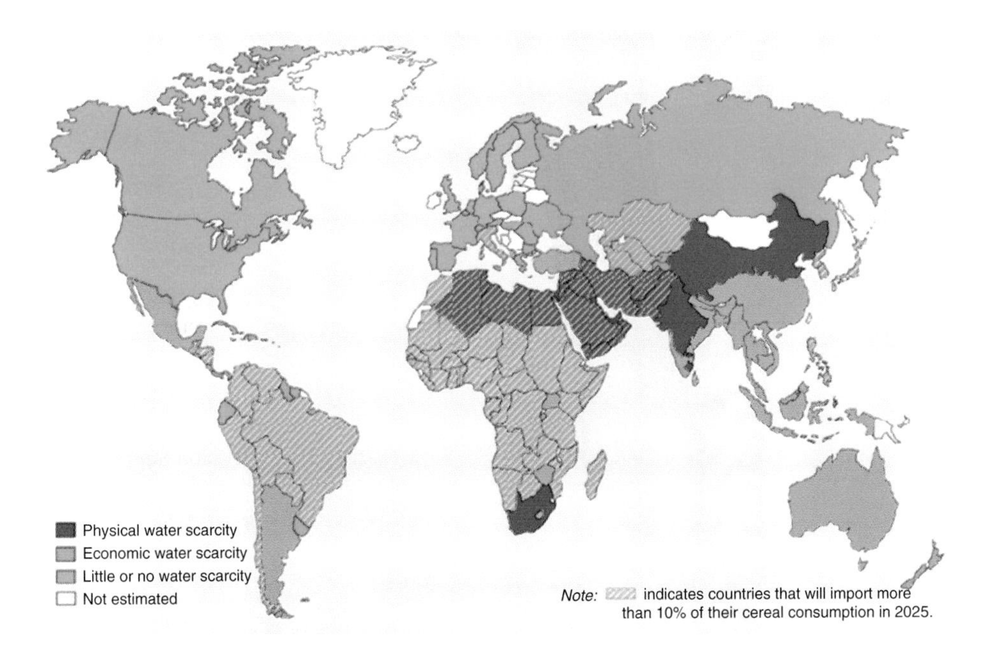

Plate 2 Projected water scarcity in 2025, International Water Management Institute

Source: http://www.iwmi.cgiar.org/assessment/files/pdf/publications/ResearchReports/CARR1.pdf

Plate 3 Pima Road recharge basin near Tucson, Arizona. (Courtesy of Central Arizona Project)

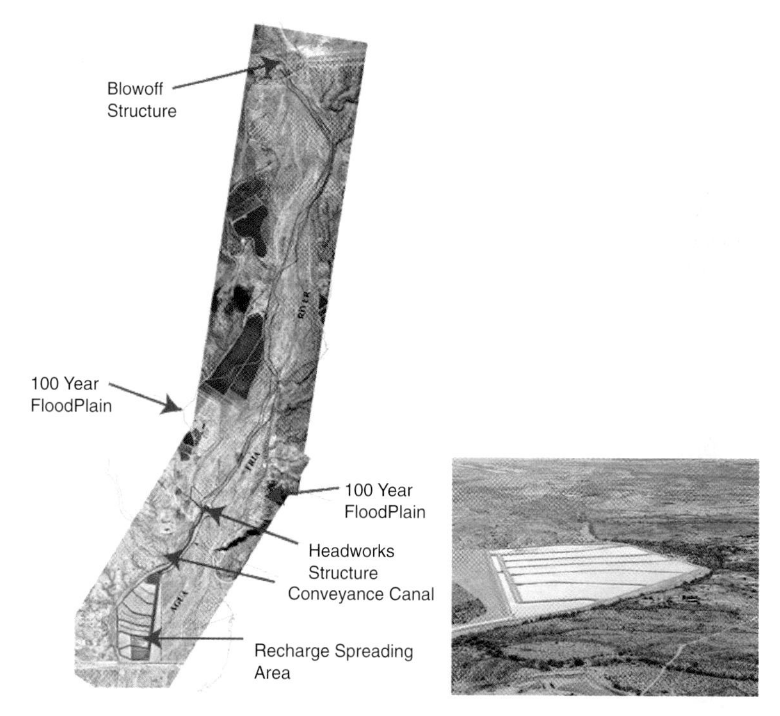

Plate 4 Agua Fria Recharge Project (AFRP) located in Agua Fria River, Peoria, Arizona. This project is located approximately four miles downstream of the New Waddell Dam (Lake Pleasant). The project was developed by the Central Arizona Water Conservation District (CAWCD) and in 2003 the City of Peoria purchased AFRP storage capacity for recharge to meet the demands of future growth as part of their water resources management goals. Two operational components include the four-mile river section used for recharge and conveyance of surface water downstream, and a constructed head structure to capture surface flow in the river and a canal to convey water downstream to the spreading basins (100 acres in area). (Courtesy of the Central Arizona Project)

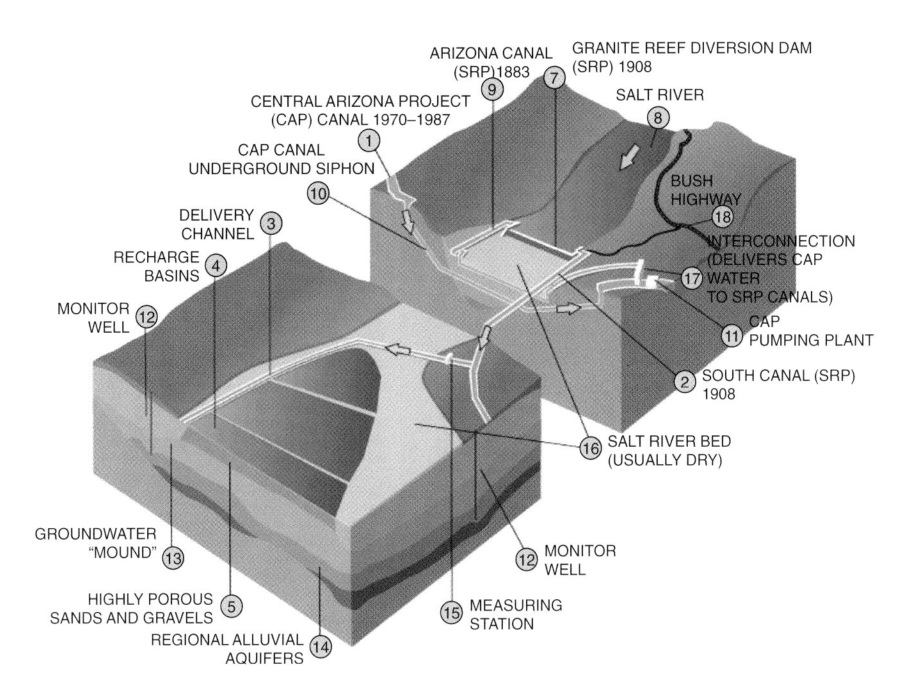

Plate 5 Granite Reef Underground Storage Project (GRUSP) operated by the Salt River Project (SRP), Arizona. This facility diverts water from the Granite Reef Diversion Dam near Phoenix, Arizona on the Salt River to seven recharge basins totalling 217 acres for the purpose of water banking. Actual recharge is approximately 100,000 ac-ft/yr with 200,000 ac-ft/yr permitted by the State of Arizona. This was the first major recharge facility in the state of Arizona and one of the largest in the US. (Courtesy of the Salt River Project)

Plate 6 Central Arizona Project aqueduct through residential area in Scottsdale, Arizona (US Bureau of Reclamation)

Plate 7 Showing rooftop drainage flowing into infiltration bed (photos by L.W. Mays)

Plate 8 Example of small neighbourhood detention basin in Scottsdale, Arizona (photo by L.W. Mays)

(a) Intake structure to detention pond

(b) Outlet structure for pond

Plate 9 Detention basin in Phoenix, Arizona showing inlet and outlet structures (photo by L.W. Mays)

Plate 10 Flood early warning system for flood control district of Maricopa County (FCDMC) (Courtesy of FCDMC)

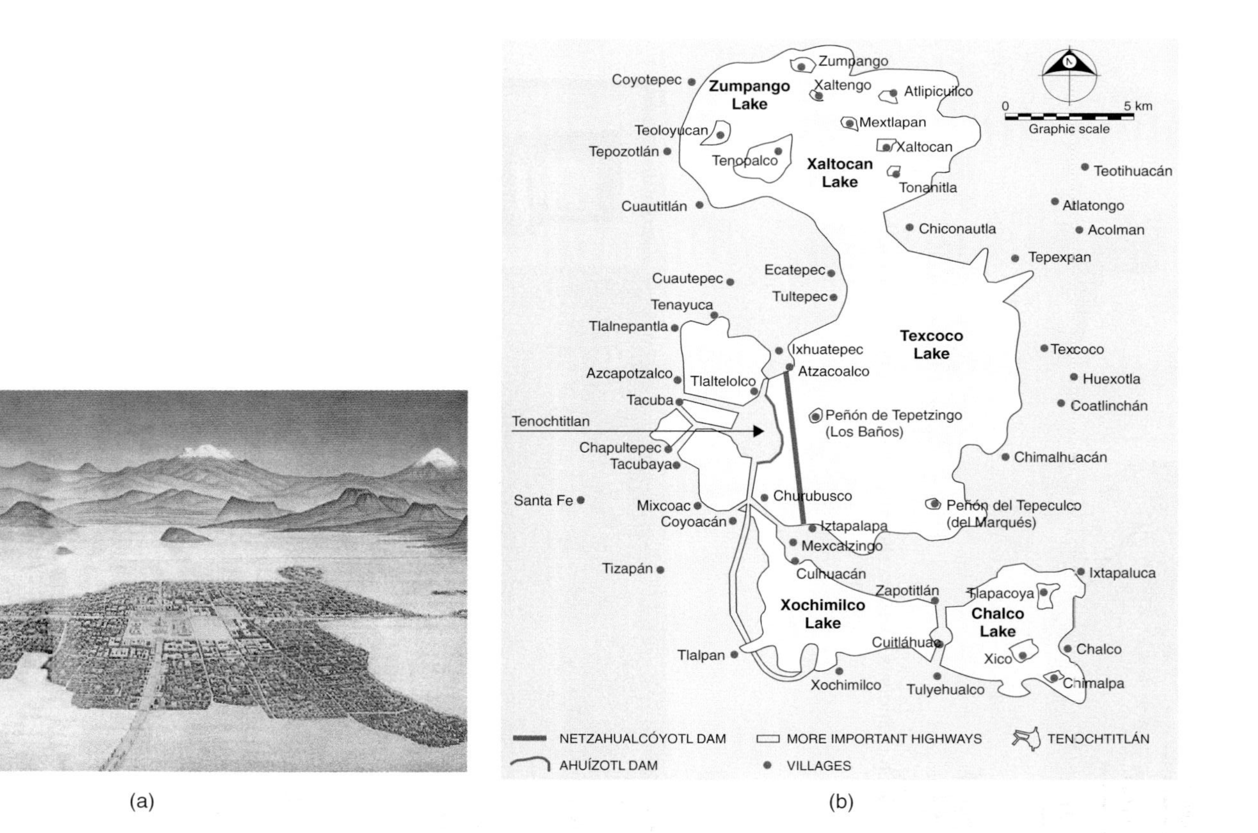

(a) (b)

Plate 11 Mexico City during antiquity **(a)** Representation of Tenochtitlan City; and **(b)** Original lakes in the Mexico Valley (From Santoyo et al., 2005)

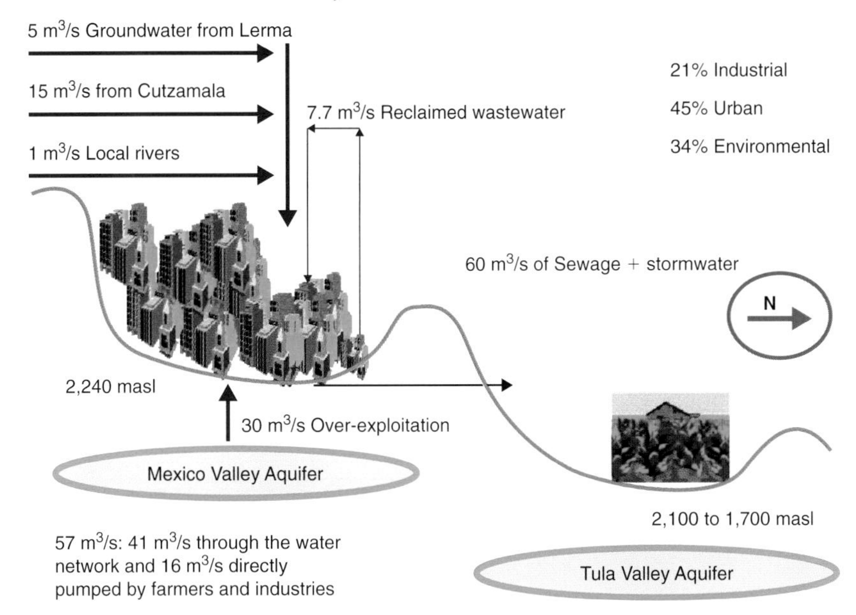

Plate 12 Water sources of Mexico City

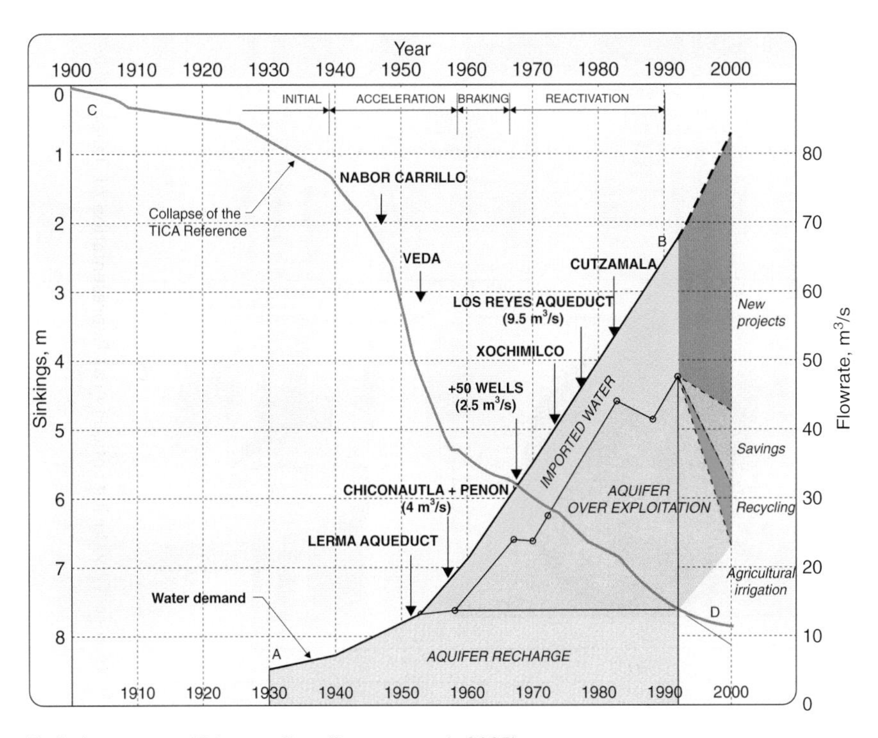

Plate 13 Sinking rates of Mexico City (Santoyo et al., 2005)

Plate 14 **(a)** Chapultepec Recreational Lake; **(b)** Birds in the Texcoco Lake; and, **(c)** Dust storm in Mexico City before the Texcoco Lake was recovered

Plate 15 El Mezquital area **(a)** with and **(b)** without wastewater for irrigation

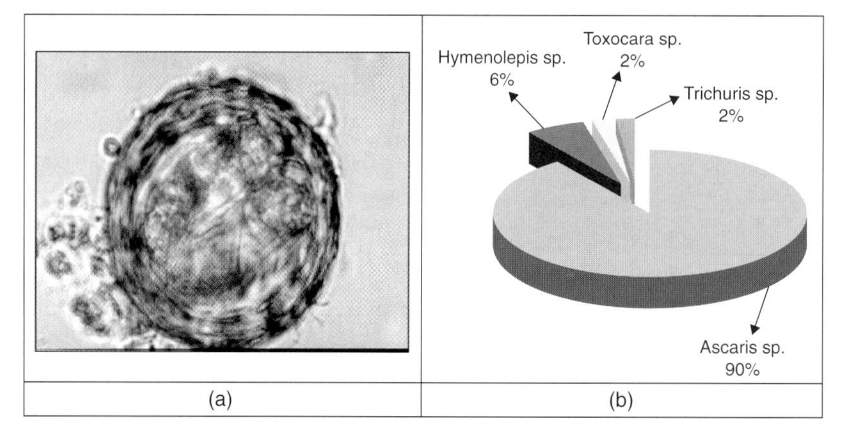

Plate 16 **(a)** *Ascaris sp.* egg and **(b)** Frequency of helminth ova genus found in Mexico City's wastewater (Jiménez et al., 2001)

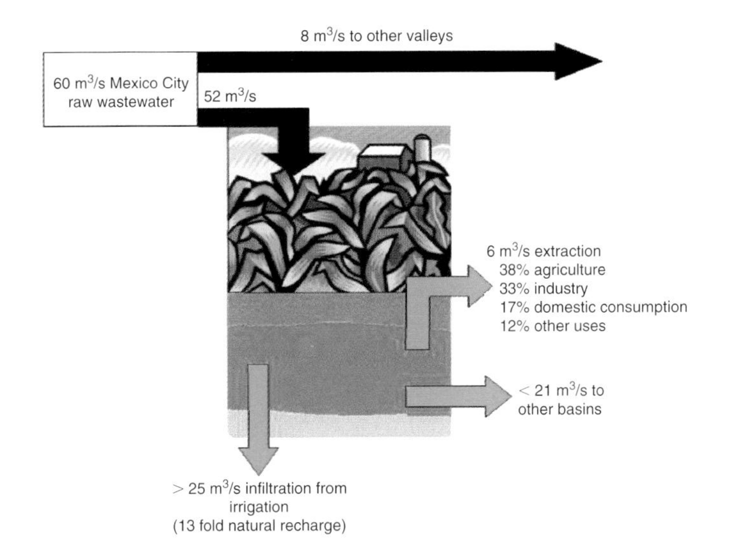

Plate 17 Water balance in the Tula Valley

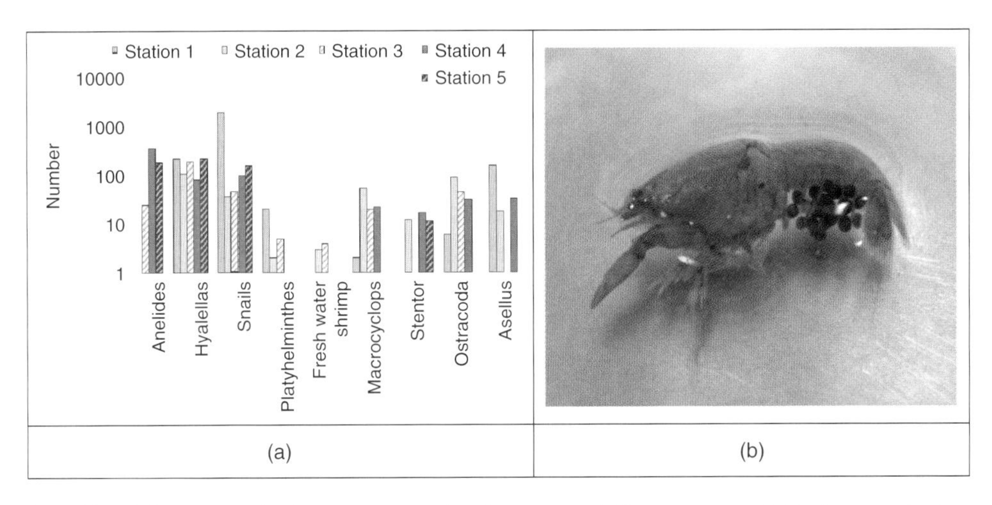

Plate 18 **(a)** Biota found in Tezontepec, a spring that appeared 30 years ago and **(b)** photograph of a Mexican shrimp named 'acocil', considered an indicator of unpolluted water

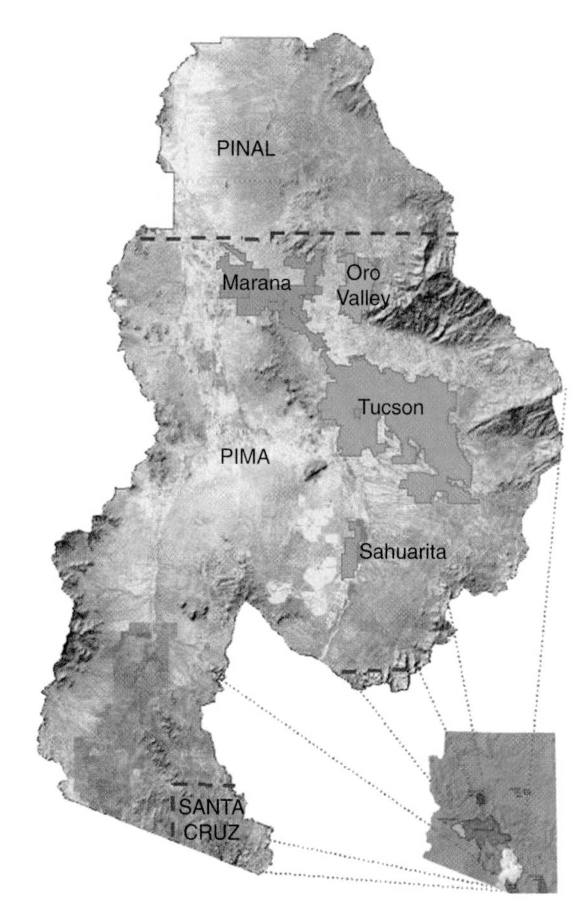

Plate 19 The Tucson Active Management Area (TAMA). The insert is provided to show the general location of the TAMA and positions of other Arizona AMAs

Source: Arizona Department of Water Resources

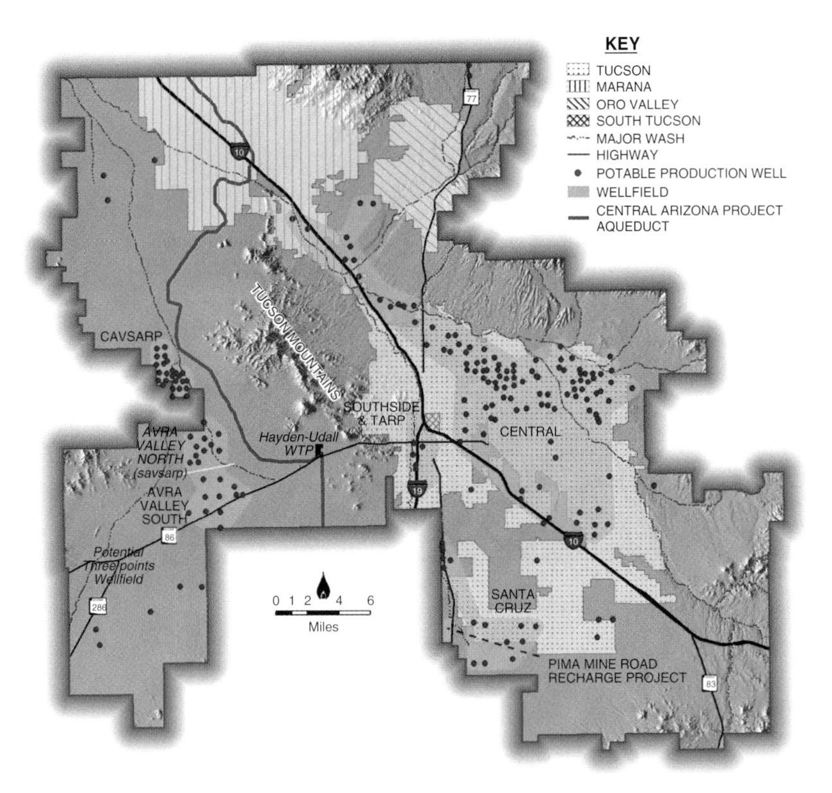

Plate 20 Major components of Tucson Water's water supply system. The map encompasses the region for long-rang integrated water supply planning in the Tucson area. (Courtesy of Tucson Water.)

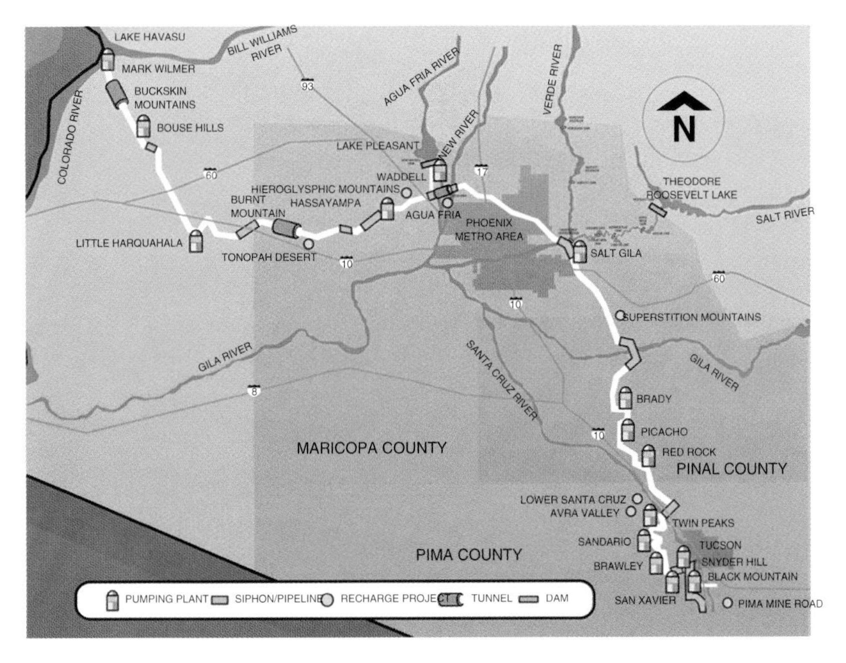

Plate 21 Layout and major features of the Central Arizona Project system. Colorado River water is mixed with water from the Agua Fria River in Lake Pleasant near Phoenix

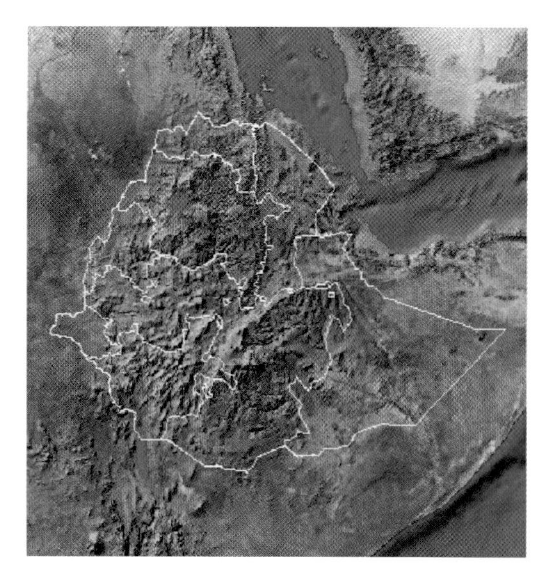

Plate 22 Ethiopia's topographic satellite map

Source: FAO

Plate 23 Landsat 7 Mosaic Cover of Awash Basin

Source: NASA

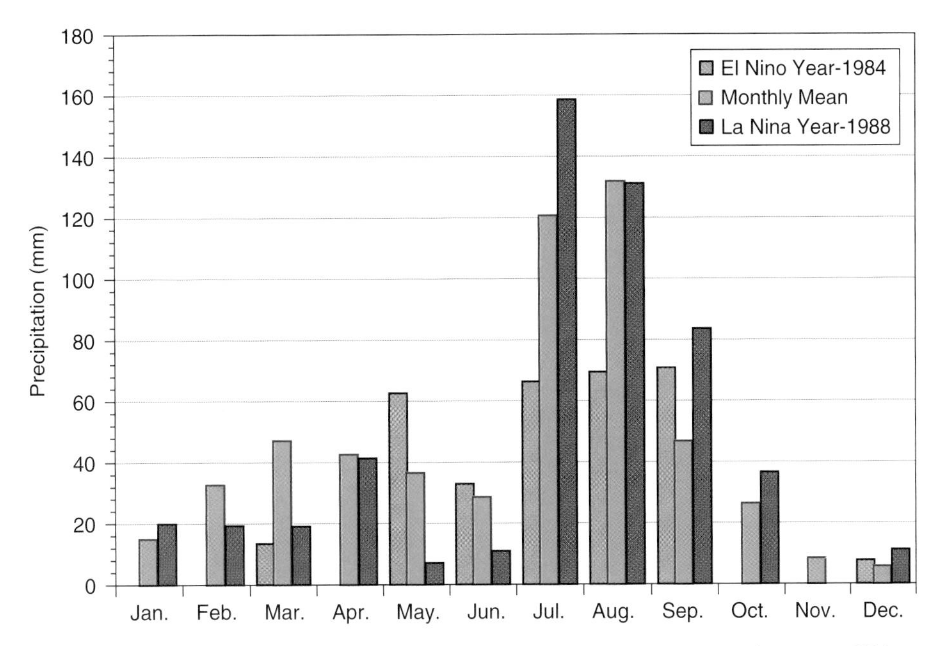

Plate 24 Effects of ENSO and LNSO events on precipitation in arid and semi-arid regions of Ethiopia

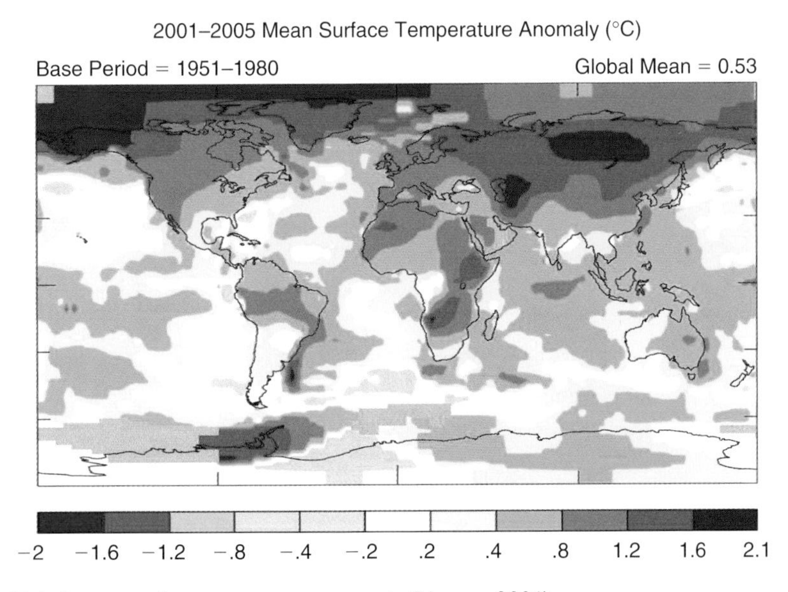

Plate 25 Global mean surface temperature anomaly (Hansen, 2006)

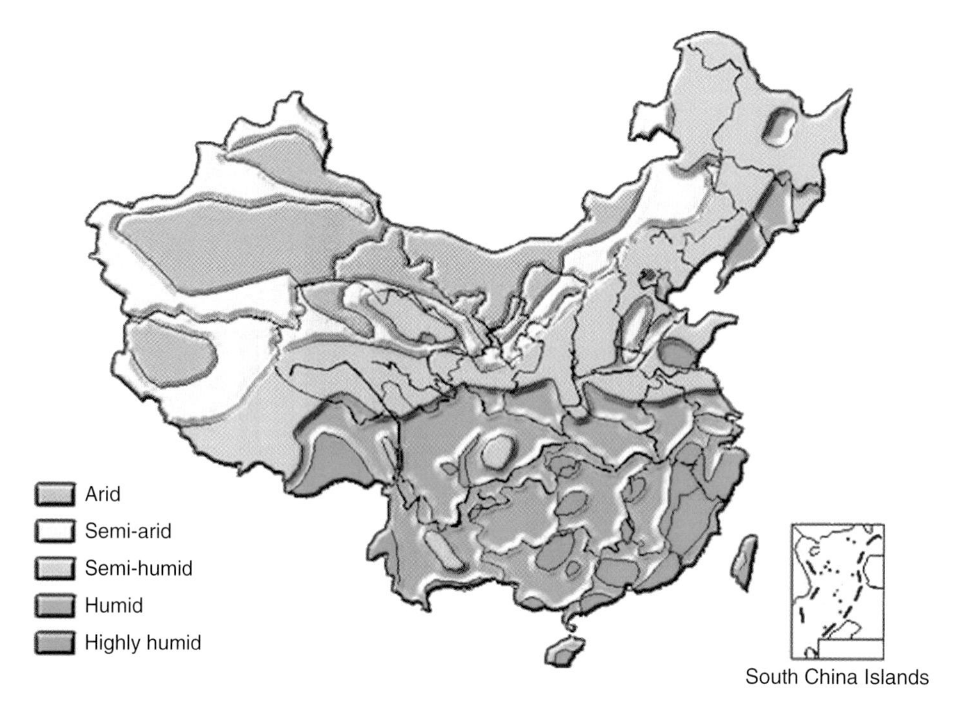

Plate 26 Distribution of water resources in China

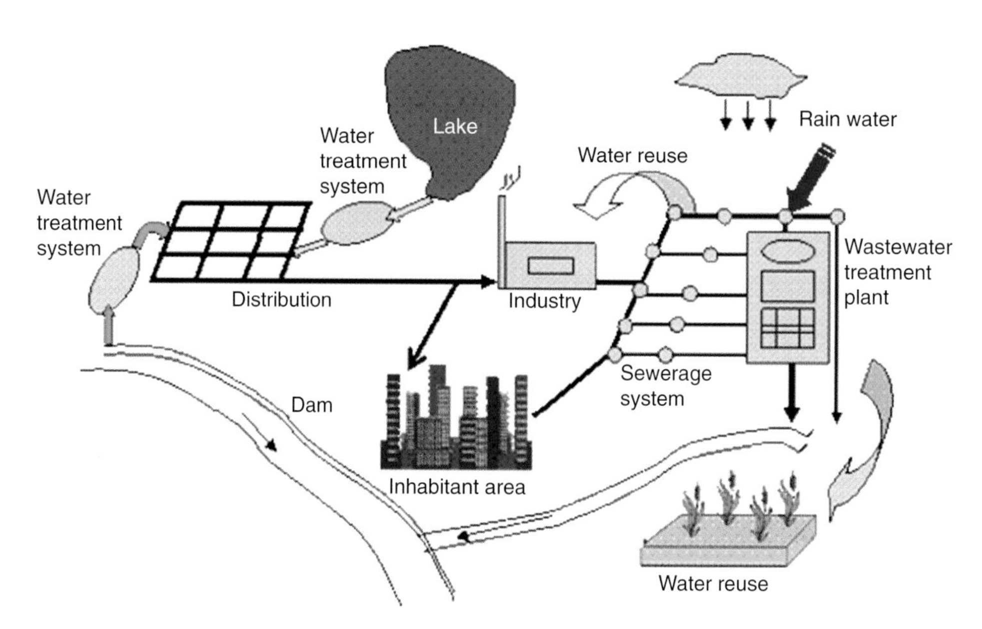

Plate 27 Urban water management system in China

Plate 28 Cairo located on both banks of the Nile River

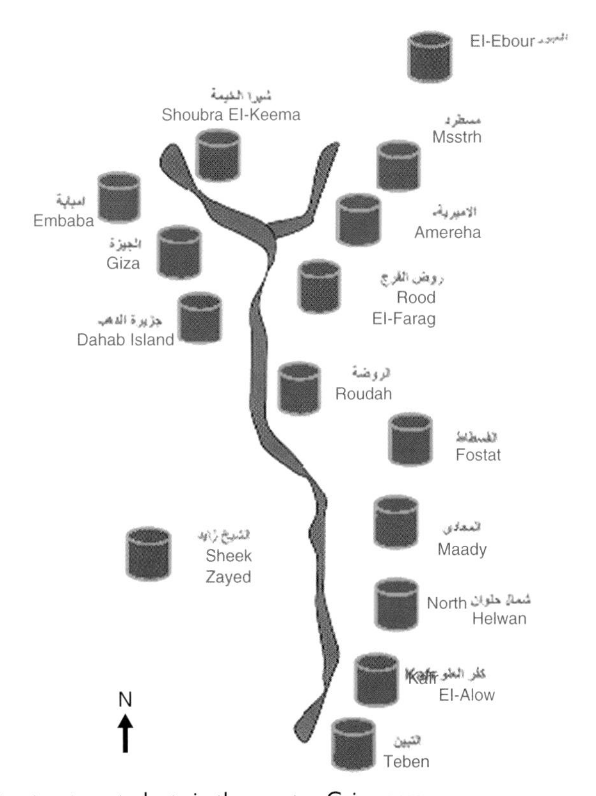

Plate 29 Clear water treatment plants in the greater Cairo area

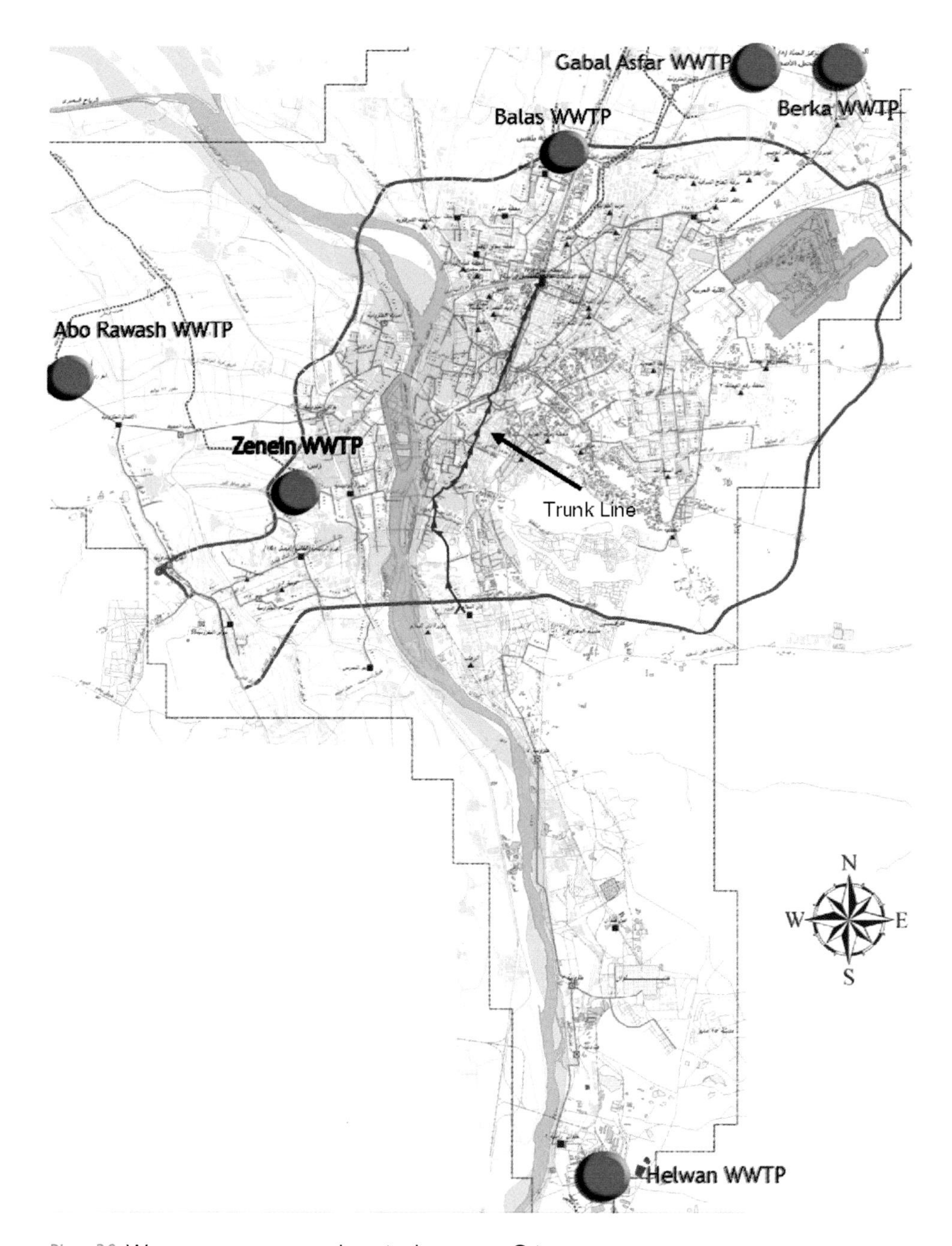

Plate 30 Wastewater treatment plants in the greater Cairo area